Total Production Maintenance

A Guide for the Printing Industry

by
Kenneth E. Rizzo

Graphic Arts Technical Foundation
PITTSBURGH

International Standard Book Number: 0-88362-199-1
Library of Congress Card Catalog Number: 97-70433

Printed in the United States of America

GATF Order No. 1543

Graphic Arts Technical Foundation
200 Deer Run Road
Sewickley, PA 15143-2328
Phone: 412/741-6860
Fax: 412/741-2311
Email: info@gatf.lm.com
Internet: http://www.gatf.lm.com

Total Production Maintenance

Contents

Acknowledgments

For their contribution of Chapter 7 on prepress maintenance, thanks to Ronald Bertolina, prepress instruction technologist, and Charles Koehler, technical consultant, both of GATF.

Thanks to Allan Fowler, Shepard Poorman Graphic Arts Center; Don Bence, MAN Roland; John Goodell, Komori America Corporation; Robert McKinney and Ron Sable, KBA-Planeta North America; and Don Goldstick, Heidelberg USA for their review of the manuscript.

Thanks also to Scott Reighard, vice president of operations, Acorn Press, for use of several photos in chapter 10.

To my sons, Douglas, Craig, and Brian, for bringing more joy to my life than they realize.

To my parents, Elizabeth and Earl, for giving me confidence.

To my sister, Corinne, for being such a good friend.

To my friends and mentors, Les, Jim, and Bob, for their support, friendship, and the knowledge that they willingly shared with me.

To my wife, Margaret, for all her love and support.

Publisher's Foreword

GATF is pleased to offer the printing industry a comprehensive, systematic approach to a business fact of life—equipment maintenance.

Total Production Maintenance: A Guide for the Printing Industry is the result of two years of development by GATF senior technical consultant Kenneth Rizzo, who adapted the well-known Total Productive Maintenance (TPM) philosophy of Seiichi Nakajima specifically for the printing industry. *Total Production Maintenance* brings TPM theory to the practical level of the printing shop floor.

The preventive maintenance (PM) movement began in the 1950s as an innovation by U.S. manufacturers in response to the "fix it when it breaks" attitude of the previous era. PM established maintenance functions as a distinct part of the industrial process. By the 1960s the theory had evolved to *productive* maintenance, also encompassing economic efficiency in plant design.

Nakajima, a Japanese mechanical engineer and management consultant, began studying U.S.-style preventive maintenance in 1950. After years of observation, he combined the Japanese concepts of total quality control and total employee involvement with those of productive maintenance. He introduced TPM in Japan in 1971, and published his first TPM book in 1984 (in Japanese only). He launched TPM in the United States in 1987 at conferences in Cincinnati and Pittsburgh. A year later, Productivity Press of Cambridge, Massachusetts, published *Introduction to TPM,* the first English-language book on the subject. It is still in print, and we recommend it highly.

Kenneth Rizzo's book *Total Production Maintenance* is yet another milestone in the maintenance movement. Now printers have a guidebook, complete with checklists and other resources, to help them implement a formal program for achieving optimum equipment effectiveness. We are proud to offer it to you, and we welcome your comments, innovations, and suggestions for improving future editions.

George H. Ryan
President
Graphic Arts Technical Foundation

Preface

Throughout my career, from press operating through various management roles, I have had to overcome numerous technical and operational problems. Everytime I was solving a problem or fighting a fire, the technical system was inoperative or non-productive. When the press had stopped, process throughput had actually stopped as well, resulting in limited cash flow. In other words, if the jobs aren't delivered on time you don't get paid.

Most printers have realized that to stay competitive they must receive the very best their process can produce in quality and productivity. Printers also understand that to be able to maximize their process, they must optimize and maintain all the components of the process. *Total Production Maintenance* focuses on optimizing and then maintaining effectiveness of the graphic arts processes.

Having brand new equipment and the best materials that money can buy are no guarantee that a printing company will produce a consistently-high-quality product. To help the printer ensure production of quality jobs and meeting customer expectations, there must be a series of operational systems in place. However, quality expectations of the customers today are at a high level, when comparing their suppliers. Customers already expect consistent quality from their suppliers; in other words, quality is a given. What the customers are now requiring is quick turnaround of their orders when rating and sometimes certifying their vendors.

World competition is putting great pressure on all printers to shorten their turnaround production time, control costs, and still maintain quality on all jobs. The name of the game today for printers is quick throughput. *Total Production Maintenance* focuses on the entire set of graphics arts processes and equipment. TPM is a process maintenance system that strives to optimize and maintain the graphic arts technical system, as well as accelerate plant production throughput. Acceleration of quality production throughput will greatly increase cash flow and the printer's chances for survival. This book is intended to help printers develop their own production maintenance system.

Total Production Maintenance's solid foundation is based on a series of commonsense measurement, operational, and continuous improvement systems used in most industries throughout the world. The first major step is to identify production interruptions and bottlenecks, and then begin the systematic elimination of the bottlenecks with the application of a series of continuous improvement tools. The system tools include statistical process control (SPC), single-minute exchange of die (SMED), total productive maintenance (TPM), the ISO 9000 quality operational standards, and education, skills, and knowledge of the graphic arts technical system process.

The chapters of this book follow this sequence, designed to help establish an effective Total Production Maintenance system.

1. **Introduction to Total Productive Maintenance**
   Maximize all the equipment of the process, then maintain it
2. **Recognizing Production Workflow Bottlenecks**
   Recognize where process interruptions and stops are occurring is the first step in accelerating production throughput
3. **Statistical Process Control Tools**
   Continuously monitor each element in the process
4. **ISO 9000**
   An operations tool that forces the printer to become disciplined
5. **Quality Assurance of Print Materials**
   Control of a process is achieved when the materials input is controlled
6. **Technical Systems Control**
   Proper equipment operation is determined through effective auditing and verification testing
7. **Total Prepress Maintenance**
   Make sure you have the best information (digital and analog) to work with on your press

8. **Optimizing the Press**
   The press must be tuned up, have predetermined maximum capabilities, and accurate fingerprints
9. **Quick-Response Makeready**
   Minimize the time required to go from the last good sheet of one job to the first good sheet of the next job
10. **Equipment Maintenance**
   Either the printer controls the schedule for maintaining the equipment or the equipment will

The printer must always remember—in the jungle, the tiger starves last!

Kenneth E. Rizzo
March 1997

1 Introduction

Total productive maintenance (TPM) was developed and introduced in Japan by mechanical engineer Seiichi Nakajima. Nakajima developed total productive maintenance by integrating the concepts of preventive maintenance (PM) and total quality management (TQM), focusing on eliminating equipment losses and an ultimate goal of zero breakdowns. TPM looks at two types of equipment losses: time loss and quality loss.

The basic concepts of total quality management (TQM) have been adapted by many companies in the graphic communications industry over the past decade. Many printers had difficulty and varying degrees of success with adopting those concepts. Today, however, the two biggest concerns that printers still have are equipment changeover (makeready) downtime and abnormal equipment operations downtime. New technology has greatly improved operating efficiency of graphic communications equipment. However, effective systems control and maintenance is still required to maximize both new and older types of equipment. The most effective system for overcoming the various types of downtime is total production maintenance.

The Six Big Losses

The TPM system first addresses six operational and mechanical losses that typically occur on graphic communications equipment:
1. Equipment failure/breakdown
2. Equipment setup and adjustment (makeready)
3. Equipment idling and minor stops
4. Reduced running speeds
5. Defective products
6. Reduced equipment yield (start-up losses)

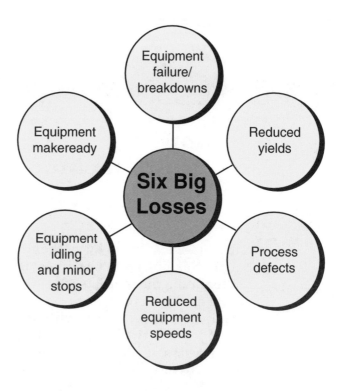

Breaking down the typical equipment downtime losses into individual components allows greater focus to start eliminating these losses. The key to TPM is eliminating both operational and maintenance causes of the six big losses.

Equipment Failure and Breakdown

There are two basic types of equipment breakdowns, sporadic breakdowns and chronic breakdowns. Sporadic breakdowns are very sudden and unexpected, often shutting down a piece of equipment for longer periods of time. Sporadic breakdowns are usually the result of deterioration of the equipment's mechanical and electrical operating components. Restoring equipment to its normal operating condition is the primary solution to sporadic breakdowns.

Chronic breakdowns and loss may only result in small amounts of lost time per occurrence, but are very frequent. Chronic breakdowns are the result of defects in equipment, tools, materials, and operating methods. The following are characteristic of chronic breakdowns:

- Have hidden causes
- Often have more than one cause
- Have a high frequency of occurrence
- Result in negligible lost time per occurrence

- Are very difficult to quantify
- Can be quickly solved, restoring equipment to operation
- Are accepted by management and production staff as a variable of the process

Overcoming chronic breakdowns and loss requires company-wide teamwork, brainstorming, and innovation. Many printers take some degree of corrective action to minimize sporadic breakdowns and losses. However, those same printers normally take very little action to solve chronic problems.

Equipment Setup and Adjustment: Makeready

With shorter run lengths, printers are experiencing more and more press makereadies. Press or equipment makeready is the time lost when changing the press over from printing the last good sheet of one job to printing the first good sheet of the next job. The term used is "last good, first good."

The most effective system for focusing on makeready was developed in Japan by Shigeo Shingo, called single-minute exchange of die (SMED). SMED focuses on two major principles: converting internal time (time when the press is stopped) to external time (time when the press is in running operation), and eliminating adjustments required to match job specifications and complete makeready. If applying SMED techniques can help reduce an average press makeready from two hours to one hour, then sixty minutes of latent or hidden time was exposed and removed.

Equipment Idling and Minor Stops

A minor stop occurs when a machine's production is interrupted by a slight malfunction or material abnormality. Examples include bad loads, feeder trips, changing loads (feeder and delivery), folder or sheeter jams, cleaning plates, cleaning blankets, cleaning dampening systems, cleaning dirty sensors, and replacing broken feedboard tapes. These temporary stops are not really breakdowns. Returning the press to normal production is easily achievable by replacing materials or resetting press components. Minor stops are normally easily remedied, but can greatly impede effective equipment operation. The main reason minor stops are overlooked by management is that they are difficult to quantify. The extent that minor stops impact equipment operations usually remains unknown. Eliminating minor stops is essential with the increased automated technology on today's presses such as automatic plate changers, automatic blanket

washers, automatic roller washers, computer-controlled register and inking, plate scanning, digital ink profile input, and
closed-loop densitometer or spectrophotometer systems. Closely
focusing on operating conditions is the only way to eliminate
the defects and abnormalities that cause minor stops.

**Reduced
Equipment
Speeds**

Slower equipment speed is a loss because a machine is not
operating at its designed speed. Slower equipment speeds
are defined as the difference between the manufacturer's
rated equipment speed and the actual speed at which the
equipment is running during normal production. The main
goal for the printer is to start bringing the actual production
speeds closer to the rated speeds until it has been reached.
There are numerous reasons that presses are run slower
than rated speeds:
• Marking of the sheets
• Slow-drying water-based coating
• Slow-curing UV
• Poor fit and register accuracy
• Difficulty delivering sheets or signatures
• Difficulty feeding sheets
• Sheet coating picking
• Mechanical press problems
• Fear of wearing out the equipment
• Ineffective monitoring of quality

Increasing the press speed may reveal hidden problems, enabling more effective problem solving.

Defective Products

Defective products continue to be a nagging loss. Losses due to defective final products include extensive inspection staff, production bottlenecks in the inspection area, time required to handle non-conforming products, job reruns, and production bottlenecks due to work-in-process defects. Defective products must be treated as the loss that it is and focused on for elimination.

Reduced Equipment Yield

Startup loss is the time when, after makeready is complete, production sheets/signatures are being counted at a reduced speed. At this time, register adjustments and color tweaking are still being conducted to stabilize the production run. Elements influencing startup loss include equipment conditions, adequacy of materials, established operating procedures, and the knowledge or skill level of operators. Startup losses are usually rather hidden, but the amount of these losses can be frequently high. Startup losses have been basically accepted as another variable of the process, thus most printers maintain little focus on their elimination.

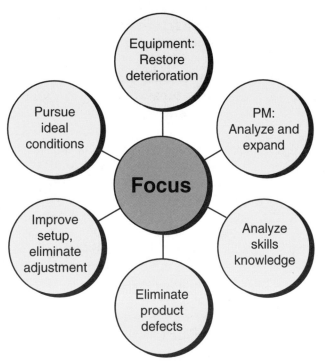

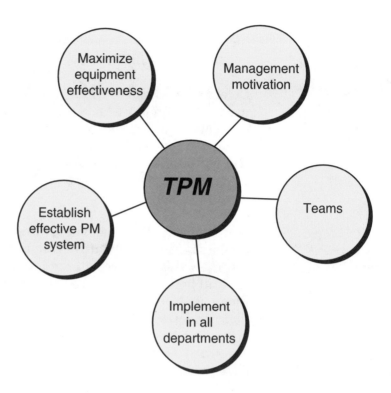

Poor operation control, non-existent training and education programs, and inconsistent operator techniques (coupled with a fix-it-when-it-breaks maintenance system) are the usual causes of the six big losses.

Implement Total Productive Maintenance

This is the basic structure for developing a TPM implementation program:

1. Maximize equipment effectiveness
2. Establish a thorough and effective equipment preventive maintenance program
3. Implement TPM in all departments
4. Establish continuous improvement operation teams
5. Implement total production quality management and motivation throughout the organization

The strategy for developing and implementing total production maintenance in plant production includes the following areas:

- **Equipment restoration and ending deterioration:** Restore equipment to original design and operating specifications. Eliminate operational and environmental causes of equipment deterioration.

- **Pursuit of achieving ideal conditions:**
  Continuously improve operations to surpass recommended industry and manufacturer conditions.
- **Improving makeready and eliminating adjustments:**
  Implement a quick-makeready improvement team program on all production equipment. The single-minute exchange of die system has been the most effective system for achieving quick changeover (makeready).
- **Elimination of internal and final defective products:**
  Eliminate internal production defects, such as incomplete job jacket information, films, proofs, platemaking, and materials. Eliminate final printing and finishing product defects, including setup waste, image fit and misregister, scumming, ink setoff, poor ink and coating rub, and bad side guide control.
- **Expansion of operator and maintenance skills:**
  Seek the equipment manufacturer's assistance to expand maintenance staff and operator skills for effective autonomous and preventive maintenance programs. Acquire the National Skills Standards for the printing industry (available from GATF) to begin development of a realistic and effective training and education program.
- **Preventive maintenance analysis:**
  Develop a team environment to focus on an effective preventive maintenance program. The team must include all levels of the organization: top and middle management, support staff, equipment operators on all shifts, and manufacturer's technical support staff. The goal is zero breakdowns.

Proactively implementing TPM throughout the operation and production areas will be the best way to eliminate the six big equipment losses.

Maximizing Equipment Effectiveness

Several steps are vital to maximizing equipment effectiveness.
1. Eliminate the six big losses
2. Develop an autonomous maintenance program
3. Develop a scheduled maintenance program for the maintenance department
4. Increase skills of operations and maintenance departments
5. Initialize equipment management program

Once the six big losses have been eliminated, there must be effective TPM teams in place for the next steps of maxi-

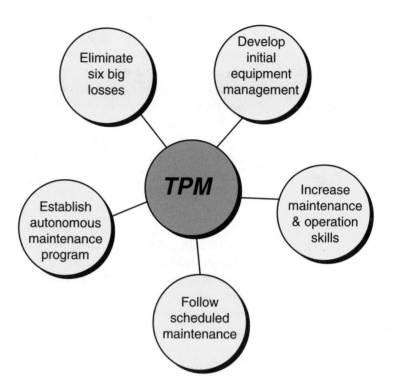

mizing equipment effectiveness to be successful. The TPM team members must be drawn from middle management, staff, equipment operators, and maintenance technicians. The keys to TPM team effectiveness are education, empowerment of the team members, and management's application of motivation. The education process must include knowledge of the TPM process, the elements of the plant's workflow, and recognition of process bottlenecks.

There are two basic maintenance activities required to improve equipment effectiveness:

1. **Maintenance operation activities**:
   Help prevent equipment breakdowns and restore equipment to required operating conditions. Establish and maintain proper equipment operations. These activities should occur over the normal cycle of preventive maintenance and production operations.

2. **Maintenance improvement activities**:
   Reduce maintenance time requirements, extend the life of the equipment, improve reliability and maintainability of equipment, establish maintenance prevention, and work toward developing maintenance-free designs.

Empower teams to implement effective TPM activities and to apply the key issues of the book *The One Minute Manager* by Kenneth Blanchard and Spencer Johnson. Two important parts of the book discuss plans for one-minute goal setting and one-minute praise.

One-Minute Goal Setting:
1. Agree on your goals
2. See what good behavior looks like
3. Write out each goal on a single sheet of paper using less than 250 words
4. Read and re-read each goal, which requires only a minute or so each time you read them
5. Take a minute every once in a while out of your workday to look at your performance
6. See whether or not your behavior matches your goal

One-Minute Praising:
(A couple times a week go into production areas. Specifically look for the good things people are doing and praise them.)
1. Tell people up front that you are going to let them know how they are doing
2. Praise people immediately
3. Tell people what they did right—be specific
4. Tell people how good you feel about what they did right, and how it helps the organization and the other people working there
5. Stop for a moment of silence to let them feel how good you feel
6. Encourage them to do more of the same
7. Shake hands or touch them in a way that makes it clear you support their success in the organization

To be able to start eliminating the six big losses a program of maximizing equipment effectiveness must be developed and implemented. This program must first include an **autonomous maintenance program.**

In **autonomous maintenance** the operator's and maintenance staff's duties and activities overlap. The equipment operator's current responsibilities of basic cleaning and lubrication expand to actually preventing breakdowns. However, autonomous maintenance also requires the equipment operators to become the eyes and ears of the maintenance department. During weekly lubrication and cleaning, equipment operators must inspect for deterioration and malfunction in all components they clean and lubricate. The scheduled cleaning and lubrication activities can be a very effective procedure in controlling equipment deterioration. While the operators are cleaning/lubricating they handle and touch each part of the equipment. The operators are looking for vibrations, unusual noises, abnormal temperatures, and loose items and components. Removing the dirt, rust, and other contaminants while lubricating the essential equipment parts will greatly slow deterioration. If any abnormalities are discovered they are recorded and forwarded to the maintenance department for prioritizing and corrective action. Program activities and tasks for the production operators should include:

1. Preventing equipment deterioration
   - Correctly operate equipment
   - Maintain basic equipment conditions with proper cleaning, lubrication, and inspection
   - Accurately maintain equipment malfunction data
   - Make proper equipment adjustments
   - Work effectively with maintenance staff to learn and prevent equipment deterioration
2. Monitoring and measuring deterioration
   - Conduct necessary daily inspections
   - Conduct periodic inspections
3. Restoring equipment
   - Perform simple and minor equipment repairs
   - Accurately report equipment breakdowns, malfunctions
   - Give maintenance staff assistance in breakdown repairs

Since equipment operators work the closest with the equipment, their activities are extremely important. Maintaining basic equipment conditions (cleaning, inspection, and lubrication) and daily monitoring and inspecting will assist the maintenance staff with equipment deterioration prevention. Program activities for maintenance staff should include:

1. Improve the maintainability of equipment
   - Improve equipment effectiveness
   - Improve equipment makeready time, downtime, equipment speeds, and production time
   - Streamline necessary maintenance, improve maintenance performance

2. Follow a scheduled maintenance program
- Acquire the discipline to adhere to structured scheduling and standards
- Proactively start to eliminate the six big losses
- Record all equipment maintenance and breakdowns clearly and concisely
- Periodically review with management to determine the effectiveness of the TPM process

3. Train and assist operators with autonomous maintenance
- Teach operators to set realistic standards and to conduct the necessary training

4. Additional activities of the maintenance staff
- Set maintenance standards
- Keep maintenance records
- Audit effectiveness of maintenance program
- Research and improve maintenance technology
- Work closely with equipment manufacturers and designers

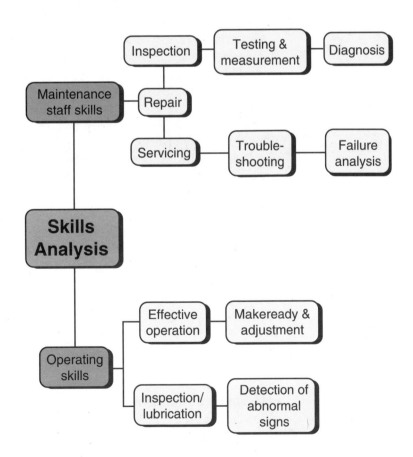

Increased Knowledge and Skills

An effective and verifiable training process must be implemented to increase skills and knowledge of both the equipment operators and maintenance staff. To eliminate the six big losses, operator and maintenance staff duties, responsibilities, and focus must expand and overlap. An effective ongoing training program is the key to developing knowledge and skills. The initial training must be thorough and its effectiveness must be verifiable.

Small group or team development and implementation of effective documented procedures for equipment operation, makeready, and maintenance activities is essential for eliminating the six big losses. Graphic communications educational institutions and equipment manufacturers are the leading resources for education and skills development. To know if an organization provides a good education program, check into the following areas:
- Is the organization acknowledged by the industry?
- Does the organization have experienced educators on staff?
- Do the education and training programs have testing methods to verify effectiveness of education and training?
- Does the training encompass both knowledge acquisition and skills mastering?

Skills analysis can be done with the help of the National Skills Standards for the Graphic Arts (for information, call GATF at 412/741-6860). There are also national skills standards for maintenance staff (contact the National Skills Standards Board at 207/985-9898).

Initial Equipment Management

Initial equipment management starts with close consultation between the equipment manufacturer, plant engineering, and maintenance staff during the equipment's design and manufacturing. The consultations must include developing an effective program for optimum equipment installation, minimizing and preventing equipment deterioration, and extending the equipment's life. The operators and maintenance staff must be properly trained by the manufacturer in all areas of operation and maintenance. Periodic checks to document the equipment conditions and capabilities, starting at the equipment's installation, must be established and performed. For presses, run a press analysis (with a certifiable test form) every six months starting at delivery.

Comparisons of the tests will quickly tell if the press is maintaining process capabilities.

Summary

The guides in the following chapters will help you eliminate the six big losses, develop autonomous and scheduled maintenance programs, increase the skill level of operators and maintenance staff, manage equipment more effectively, and decrease makeready time. Once you have established a complete TPM program, you will be able to maximize output from your press.

2 Recognizing Production Workflow Bottlenecks

One of the most frustrating problems that printers face every day is production bottlenecks. Bottlenecks or production throughput interruptions are isolated or temporary slow-downs and stops in the production process. A bottleneck usually means that a process component or piece of equipment cannot produce fast enough to meet demands. Many people view production bottleneck loss as the cost of lost time on a specific piece of equipment, which may cost the printer a few hundreds of dollars per hour to operate. The real lost time must include the effect of the stop on total plant production throughput. In actuality, the lost production throughput costs the printer many thousands of dollars per hour. Production bottlenecks slow production throughput, delivery to customers becomes unreliable, and cash flow is limited.

The variety of causes of these interruptions fall mainly under two categories, the technical system and the quality system. The technical system includes all the equipment, materials, tooling, and instruments that are required for the printer to produce the printed product. The quality system provides the controls and documentation for a proper technical system and includes company policies, defined quality, planning procedures, operating techniques, measurement techniques, training, and work instructions.

There are two types of bottlenecks, chronic and sporadic. Chronic production interruptions or bottlenecks will happen regularly and are many times accepted by printers as part of the production process. Chronic bottleneck causes can include late or incomplete job information on the job jackets, abnormal equipment conditions, older equipment, slower equipment, specific customer demands, skill and knowledge levels of overall staff, production workflow planning/scheduling, poor plant environmental controls, and typical uncontrolled technical system and material variables to name a few. However, the vast majority of chronic production bottleneck causes are basically hidden and at times difficult to detect. The main reason that causes of chronic loss appear hidden is that chronic bottlenecks usually waste smaller amounts of time (usually from a few minutes to an hour) per incident. Chronic bottlenecks are often easy to correct, but frequently there is little focus to eliminate the root causes. However, chronic bottlenecks occur with a high degree of frequency and over an annual period their total lost time can be extremely high.

Sporadic bottlenecks, on the other hand, are interruptions and stoppages that happen very suddenly, but with a low degree of frequency. Sporadic bottlenecks are losses that will waste large amounts of time (usually hours and days) and can cause virtual shutdowns of plants or processes. Since sporadic bottlenecks and interruptions are very conspicuous, their causes are usually obvious. Typical examples of sporadic bottlenecks include equipment, electrical, and mechanical failures or breakdowns, late raw material deliveries, defective raw materials discovered at the machine, incorrect and missing job information/specifications, and rejected work-in-process or final product.

When developing a step-by-step process to eliminate production bottlenecks and waste the printer must focus on several key issues:

1. Management must form and proactively support effective autonomous teams to focus on eliminating production interruptions and bottlenecks
2. Use SPC tools to enable increased productivity, to measure quality and productivity, and to identify the extent of production throughput interruptions
3. Identify the location of production bottlenecks
4. Identify the types of bottlenecks (chronic or sporadic)
5. Identify the causes of the bottlenecks (technical or quality)
6. Develop and implement strategies and solutions to eliminate the root causes of production bottlenecks

Typical Print Production Bottlenecks

There are a variety of causes of production interruptions and bottlenecks that are common to the graphic arts industry. The list of bottlenecks and production interruptions can be lengthy, but there are common ones that continually plague graphic arts production throughput.

As job information travels through a printing facility, maintenance affects every department.

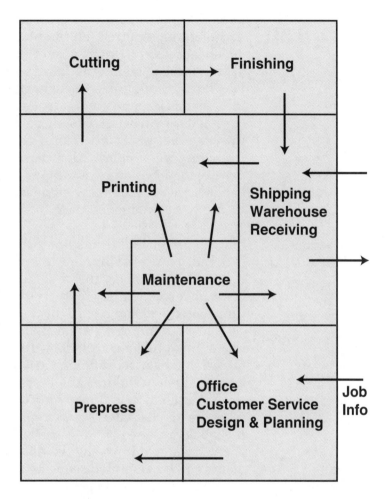

Production Planning and Control Bottlenecks

Job Planning and Documented Specifics

Frequent production bottlenecks include poor contract review procedures resulting in job jackets with incorrect, conflicting, and missing job information. The result of job jackets not being press-ready is lost materials and production time (waste). Typical qualitative job specification inconsistencies include wrong type/quantity of materials ordered, wrong copy, and wrong processes or machine specified. When a job is planned for production, a number of technical system constraints must be taken into consideration. To help overcome production bottlenecks from poor planning and information it is imperative that effective contract review procedures be established. First, people responsible for planning and documenting the jobs must be provided with ongoing training to better understand and continuously improve their job and their knowledge of the graphic arts process.

Second, establish effective procedures for job planning and documenting required information, including not scribbling on job jackets and proofs when changing job requirements. Third, develop job information and procedural checklists for job planners will greatly help to standardize tasks, minimize inconsistencies, and eliminate mistakes. (See Appendix for checklists of various production stages.) Fourth, computerize the job jacket information form and checklists to print out fresh ones when information changes are required. Fifth, a pre-production quality assurance process must be established to detect incorrect and missing job information before it enters the production cycle.

Production Scheduling

Smooth production throughput without delays or bottlenecks is the ultimate goal for all printers. The person or persons responsible for production scheduling are commonly caught between a rock (satisfying their external customers with on-time deliveries) and a hard place (satisfying their internal customers, minimizing fire drills). Some common production scheduling problems include waiting on proofs, late raw material deliveries, running partial quantities and lifting jobs, equipment failures and breakdowns, scheduling preventive maintenance, and reserving job time in the production schedule for customer service representatives (CSRs) before the job is press-ready.

Pro-actively addressing two specific production control elements will be helpful in reducing bottlenecks. Concentrate efforts first on adhering to scheduled preventive maintenance requirements. Establishing and consistently conducting effective preventive maintenance will greatly reduce unscheduled equipment downtime and accelerate production throughput. The other element to address is when to add jobs to the production schedule. Traditionally, jobs are put into the schedule so everyone can see what is coming up and how busy they are. However, the production planning department may tend to feel that their jobs have reserved parking spaces in the schedule. When a reserved parking mind-set occurs there may not be effective follow-up on proof approvals and raw materials deliveries. The results can be frequent production schedule changes, pulling jobs prior to completion, and idle equipment waiting for premakeready completion. One way to overcome this problem is to have a staging location in the schedule board or digital production spreadsheet for all jobs not ready for production. The job then does not go into the production schedule until documents indicate that all information is approved and materials are press-ready and staged. A brainstorming team environment will be needed to give the production planning people the skills and tools for effective production planning and scheduling.

Prepress Bottlenecks

Customer Supplied Products: *Preflighting*

Major headaches for printers are prepress materials and information supplied by the customer. There are no guarantees that films, proofs, and digital files supplied by the customer are correct or compatible with the printer's process. The best way to overcome supplied materials problems is to conduct a pre-inspection of the supplied materials.

The newest form of prepress quality inspection is preflighting. Preflighting (pre-production quality control inspection) of all digital files is an absolute necessity. The goal of preflighting is to discover any missing information, items, and errors in customer-supplied files. Always communicate to the customer what errors were discovered, then determine who should take the corrective action.

Inspection and proofing of all conventional prepress materials supplied by the customer is essential to prevent production bottlenecks. Typical elements checked include physical condition of films and flats, original films versus duplicated film, correct copy, color breaks, internal image fit, and matching proof to proof (customer's versus printer's). Inspection of conventional prepress components provided by the customer is a form of conventional preflighting.

Preflighting all incoming job information and conventional and digital prepress components today is the best strategy for the printer to prevent production bottlenecks. The printer should first establish specific preflight checklists and procedures. Specific people in the organization should be designated and trained to become preflight technicians. People who make preflighting effective have a proactive attitude, are patient, are thorough when completing tasks, have the ability to follow established procedures, understand the importance of documenting information and maintaining records, and have skill and knowledge of the process.

Plate Remakes

Plate remakes are a work-in-process product rejection. Plate remakes continually cause production bottlenecks in both the prepress and press areas. The prepress team should focus on the following areas when addressing plate remakes:

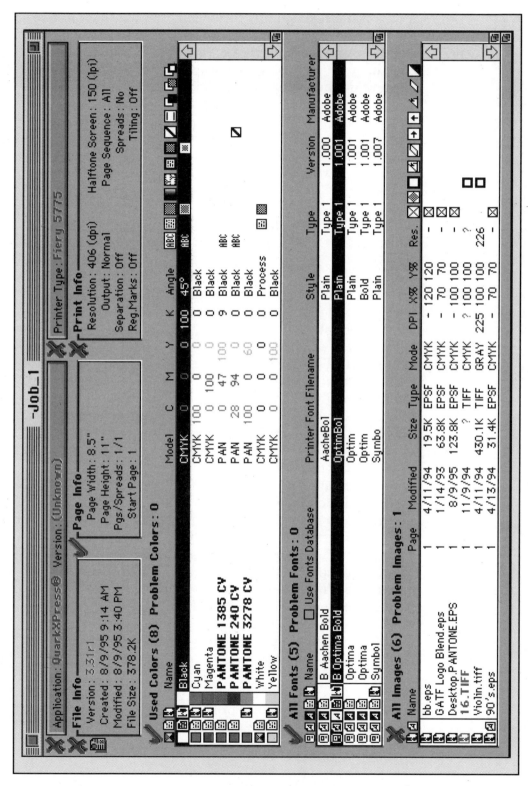

FlightCheck (by Markzware), a preflighting software system, creates a computerized report of all file information, including picture, font, and color usage, and any missing information.

- Identify and document amounts and reasons for plate remakes
- Expand knowledge and training of planning and prepress staff
- Evaluate, improve, and document structured operating procedures
- Establish and standardize effective operating techniques and procedures
- Conduct periodic equipment measurement and accuracy testing
- Develop and implement effective equipment maintenance
- Audit review procedures to evaluate effectiveness of improvement efforts

Press Bottlenecks

Excessive Dot Gain

Dot gain is both the physical and optical increase in the size of halftone dots during prepress and press operations. Dot gain is a normal part of the printing process. The current industry acceptance of the level of dot gain is recognized and published in specifications such as Specifications for Web Offset Publications (SWOP) and Specifications for Non-Heatset Advertising Printing (SNAP). However, excessive dot gain levels will cause production stops. The typical production interruptions caused by excessive dot gain can include difficulties matching proofs, long makereadies, plate remakes, film remakes, and changing inks. The printer must control several basic process elements to minimize dot-gain-related production bottlenecks:

- Maintain recommended plate-to-blanket squeeze, 0.004–0.006 in. (0.10–0.15 mm) for compressible blankets and 0.002–0.004 in. (0.05–0.10 mm) for conventional blankets
- Maintain recommended impression cylinder pressure
- Properly set ink and dampening rollers, check settings a few times a week
- Maintain fountain solution temperature between 50–60°F (10–16°C)
- Control temperature on ink rollers
- Purchase inks and fountain solution based on dot gain performance and compatibility
- Purchase blankets based on dot gain and print performance
- Proof all film, customer-supplied and internally-produced
- Maintain proper and consistent film output and processing
- Maintain proper and consistent plate output, exposure, and processing

- Provide densitometric tools to measure dot gain, use spectrophotometric tools if required
- Use control targets on film, plates, and press to enable dot gain measurement
- Provide staff with education in the process and dot gain acceptance
- Minimize hard/glazed rollers, apply an effective roller-cleaning compound to rollers
- Establish dot gain characteristics through a press optimization process using a certified press test form
- Establish color management procedures from job planning through press
- Establish quality and operation system procedures (ISO 9000)

Fountain Solution and Dampening Control

Dampening control still appears to be a problem in the lithographic process. Lost time is still incurred from dampening-related problems, including poor pH/conductivity control, ink bleed into the fountain solution, plate scum, difficulty printing solids with reverses, ink/water balance difficulties, running alcohol substitutes, ink roller stripping, and water banding. There are a number of things the printer should do to minimize dampening problems:

- Use fountain solution that is compatible with the inks and plates
- Maintain proper ink and dampening roller settings
- Install dampening system rollers designed to properly meter fountain solution
- Perform effective roller-wash techniques
- Minimize roller-glaze by applying an effective roller-cleaning compound to rollers
- Establish and maintain an effective preventive maintenance program
- Install a water-treatment system (e.g., reverse osmosis or LithoWater™) for consistency
- Mix appropriate fountain solution concentrations based on the supplier's recommendations
- Install a fountain-solution automix system for consistent concentration
- Autofeed the fountain solution in small amounts and maintain at mid-tank levels
- Install effective recycling filters in the fountain solution in-feed lines

- Maintain pressroom environment at 70–75°F (21–24°C), 45–55% RH
- Measure and chart pH/conductivity of water daily, change if conductivity rises 600 micromhos
- Maintain accurate calibration of pH/conductivity meters
- Provide staff with education in the process and fountain solution requirements

Hickeys

Particulate contamination (hickeys) has been a problem for printers for a long time. Hickeys are eyesores, negatively impacting the appearance of otherwise good-quality printing. The bottom line is, customers hate them and few will accept them. Battling hickeys can cause frequent press stops to clean plates, blankets, and rollers. The worst scenario is when the customer totally rejects the job. Eliminating hickeys requires the printer to establish and maintain an optimum pressroom environment, realistic materials procurement, and technology assistance.

- Maintain clean presses, ceilings, walls, air-quality control
- Maintain proper roller settings
- Establish and perform effective roller rinsing, cleaning techniques and procedures
- Use an effective and safe roller cleaning solvent
- Apply an effective roller cleaning compound to rollers as a maintenance cleaner
- Maintain effective wash-up tray blades
- Establish and maintain an effective roller preventive maintenance program
- Establish realistic and effective materials procurement procedures, including internal and external ink and paper/board materials information for performance predictability
- Purchase presses with, or retrofit to current presses, Delta dampening systems
- Install anti-hickey ink-form rollers in the first roller position
- Educate staff in the process of hickey intolerance
- Problem-solve hickeys and spots—gather physical evidence and accurate information

Common Causes

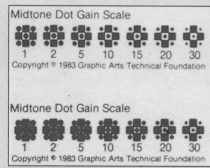

Dot gain caused by poorly-controlled ink consistency or incorrect plate exposure

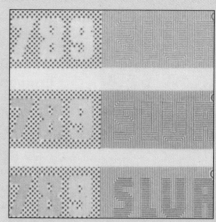

Slur caused by a loose blanket, or by ink form rollers being set too tight to plate

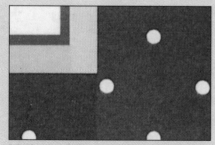

Four-color job will not fit on press, caused by inconsistent print length from unit to unit

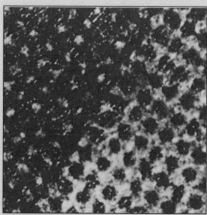

Scumming caused by insufficient moisture on the plate or weak fountain solution concentrate

Mechanical ghosting caused by insufficient ink supply on the ink form rollers to the form, or by too much water run on the plate

# of Bottlenecks

Gloss and chemical ghosting caused by a drying ink film affecting the drying of a previously-printed film

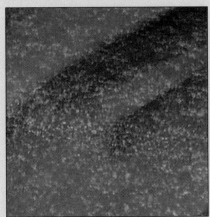

Multicolor ink trapping problems, caused by the second-down ink being tackier than the first

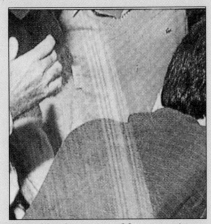

Plate blinding caused by excessive gum in the fountain solution or improper pH of fountain solution

Ink will not dry on the sheet, caused by too much acid in the fountain solution or too much water run on the press

Hickeys are caused by dirt or foreign particles (including bits of dried ink) contaminating the plate or blanket

Poor First-Pull Sheet, Register, Fit, and Color

The goal of every manufacturer is to be able to sell the very first item produced after the machine is setup. The goal of every printer when the makeready is complete is the consistent achievement of "first sheet or first product good." Due to shorter and shorter run lengths and lead times, printers today are confronted with the need for accelerating throughput with quick-response makereadies. Pursuit of achieving "first sheet good" is an integral part of the quick-response makeready process. The goal of consistently achieving first sheet good may seem unrealistic to all but a few printers. However, those few printers that view first sheet good as a realistic part of their process are usually the ones considered to be leaders in the industry. What sets this group of select printers apart from the rest of the industry is the fact that they maintain focus on their process and their goals. There are some major steps to follow when moving forward with a quick-response makeready process. These steps include focus on internal and external process components in all departments.

- Establish a proactive environment (team problem solving)
- Optimize the technical systems elements
- Implement a quality system
- Apply Shigeo Shingo's single-minute exchange of die (SMED) system

Poor Match of Press Sheet and Approved Proof

Matching press sheets to a proof has been a problem for a long time. Photomechanical and digital proofs were not really intended to be absolute matches to the job on press. Proofs were intended to be a close representation of what the printed job should look like. However, the industry and its customers many times demand almost perfect matches of the press sheet to the proof. The printer can still come consistently close to a press sheet and proof match. The printer must look at the process as a system and take a step-by-step approach to calibrating the capabilities, materials compatibilities, and accurate film output to achieve success.

- Perform a press-system optimization
  - Conduct a press-diagnostics test using a certified test form and take technical system corrective action where required
  - Perform a capability study of upper and lower control limitations of process ink production densities
  - Determine maximum process ink densities through print contrast calculations

First-Pull Register

Achieving register and image fit on the first pull is the goal of every press makeready. When first-pull register is achieved consistently, matching color specifications and checking print sheet quality can be done simultaneously. The makeready will then move along quickly, and downtime is shortened. Press technology developments such as automated plate mounting, automated blanket and roller washing, automated feeder and delivery setting, cylinder zeroing, and remote cylinder pressure setting are helping printers shorten the makeready process. For printers to attempt the goal of "first-pull register," they must concentrate on the register and pin systems. Printers frequently overlook these.

Checking Images on Plates

The first step toward achieving first-pull register is to evaluate the placement of the images on the plates. The environment in the prepress department should be clean and maintained at 70–75°F (20–24°C) and a relative humidity of 45–55%. The films and film flats must be inspected. Pay particular attention to the following areas:

❑ Make sure that the films are correctly punched and/or mounted on the carrier sheets (7-mil clear carriers for close register work) and punched in register.

❑ Examine and test the plate frame or step-and-repeat machine to determine if consistent exposure and register (a variation of less than 0.001 in. [0.025 mm]) plate to plate is being achieved.

❑ Check the plate punch for squareness and self-centering accuracy.

❑ Inspect the punch dies to make sure that they match the pins in the frame or step-and-repeat machine.

Checking the Condition of the Press

The sheet infeed, side guide, and unit-to-unit sheet transfer register must be accurate. A press test analysis using the GATF Sheetfed Test Form and the GATF Register Test Grid, which are designed to accurately determine the status of the press, should be conducted. If the status of any of these press components is unsatisfactory, corrective action must be taken to ensure that the press is operating as it was designed.

Finally, standard operating procedures (SOPs) should be developed and implemented for the changing of plates if no SOPs currently exist. Plates should be staged at printing units prior to the completion of the current job running on press. When the old plates are removed, the cylinder plate clamps should be set according to the pin system on the press.

Particular attention must be paid to the position of the gripper edge clamps: they must be placed in the correct position on every printing unit. When the new plates are mounted to the cylinder, the gripper edge plate clamps must remain in position. This should be checked in accordance with the particular pin system on the press; if the clamps do not remain in proper position, the plates will be out of register. SOPs should be developed to ensure that the plates are properly mounted and torqued up to the cylinder consistently and that the plates are not pulled off the pins or torn. The best way to develop plate mounting SOPs is to seek guidance from the manufacturer of the pin system.

The goal of consistent first-pull register is not an unrealistic one. By treating the pin system as a system, and ensuring company-wide teamwork from prepress through pressroom, this goal can be achieved.

— Establish target ink densities by calculations based on the upper and lower process ink density control limitations and maximum process ink densities provided from print contrast calculations
— Run the standardization pressrun with a certified test form to establish ink density and dot gain characteristics based on the most common inks and paper/board used on press
— Provide dot gain information to color separation for consistent output of color-corrected films
• Proof all customer-provided films and compare to proof provided to determine if film corrections are required
• Make proof on the same substrate to be printed on press
• Maintain use of standardized process inks compatible with proofing system
• Maintain press capabilities such as printing pressures and sheet transfer system
• Evaluate possible changes in print characteristics when a change in materials is to be implemented such as ink, blankets, and fountain solutions
• Maintain standard lighting viewing conditions of 5000K color temperature and 185 ($\pm$25) footcandles

Image Slur and Doubling

The worst-case scenario of dot gain is caused by image slur or doubling on the press. Image slur of 0.001 in. (0.025 mm) can be visually detected without any magnification. The dot slur causes can be either related to the mechanical operation of the press or improper materials procurement and handling. Printers agree that image slur and doubling are problems that should be corrected. However, many printers will accept the problem if they feel it is not too severe. One example of an image slur and doubling cause is mechanical image slur on double-size, drum-cylinder sheetfed presses occurring on every other sheet. Visually, the slurring will cause excessive dot gain to look the same on every sheet. The following corrective and preventive actions will greatly minimize image slur and doubling problems:
• Keep sheet-register and transfer-gripper systems properly cleaned and lubricated
• Conduct diagnostic press tests biannually as part of maintenance prediction
• Have a factory-trained technician conduct annual maintenance checkups

- Properly pack plates and blankets for recommended plate/blanket or blanket/blanket pressure squeeze
- Properly torque blankets around the cylinders
- Maintain adequate (but not excessive) impression-cylinder pressure
- Set all system rollers to manufacturer specification
- Run color bars on all jobs with star targets for slur indication

Paper Problems

When paper becomes a production interruption, it usually occurs unexpectedly on press. Most printers do not have the resources to conduct sophisticated paper tests. However, there are a number of commonsense things the printer can do to minimize paper-related downtime. To start with, effective procurement procedures will greatly minimize paper problems. This includes purchasing from paper mills through reputable paper merchants. Other procedures should include:

- Order paper/board from paper mill
- Order the right paper for the right job
- Avoid ordering generic paper; it could be rejected from the mills or other printers
- Inspect and document paper shipping containers for shipping damage
- Verify the order invoice with shipping order
- Conduct basic paper testing, such as K&N test inks for absorptivity and Dennison pick wax test for paper strength to ink tack; measure the caliper of at least five consecutive sheets from the skid or carton.
- Gather physical evidence and accurate information when a paper problem occurs
- Store paper/board in manufacturer-recommended environment of temperature and relative humidity
- Restore and maintain press equipment at optimal operating conditions
- Communicate and educate customers in paper manufacturing and characteristics

Gloss or Chemical Ghosting

Work-and-turn or work-and-tumble jobs often become a bottleneck problem for printers. One problem is the amount of time required for the ink to dry so the back side of the sheet can be printed. Another problem is that a matte ghost image can appear solid on the sheet's front side when the ink dries on the back side. Ink-solvent gases flash off the image from one side of the sheet during the drying process leaving residue on the back of the sheet, resulting in a reversed matte image appearing. There are a few things the printer can do to minimize the causes for gloss ghosting:

- Print with low-solvent inks, avoid printing with quick-set inks
- Allow a minimum of 24 hours for the first side to dry
- Overprint each pass of the job with overprint varnish or aqueous coating during first pass
- Print the heavy, solid-image side of the sheet first
- Run the job in small lifts on racking skids
- Overprint the problem sheet side with gloss varnish or coating if problem occurs
- Avoid adding gel tack reducers
- Aerate the job when possible

Mechanical Ghosting or Ink Starvation

Also known as ink starvation, mechanical ghosting is common on offset lithographic presses. Mechanical ghosting is uneven ink coverage in solids caused by different amounts of plate image around the cylinder. Mechanical ghosting can appear to be nonexistent or very subtle in some ink colors such as reds, magentas, blues, and black. With other inks, such as grays, browns, and greens, ink starvation can appear very obvious and unacceptable to both the printer and customer. Unacceptable mechanical ghosting problems can turn a highly productive press into a production bottleneck. Different inking-system designs can also aggravate potential mechanical ghosting problems. The printer should focus on both the technical system and job planning/layout to overcome or at least minimize mechanical ghosting problems.

- Properly maintain ink and dampening-system roller settings
- Implement a structured roller rotation and preventive maintenance program
- Increase pitch (amount of oscillation) of main vibrator rollers if possible
- Print ink takeoff bars at tail of sheet
- Maintain minimal amounts of fountain solution to the plates, increase strength of fountain solution and alcohol substitutes
- Install oscillating ink form rollers in at least the number three and four roller positions
- Print with an opaque ink if possible
- Print with a weaker pigmented ink to increase ink film thickness on rollers
- Angle image exposed on plates a minimum of 6°

- When purchasing a press, perform mechanical ghosting tests using the GATF mechanical ghosting form. Comparing different presses is recommended.

Press Equipment Failure and Breakdowns

Equipment failure can include both poor quality and total machine breakdown. Equipment failure does not occur when the equipment is standing there with the power off. Failures always seem to occur when hot jobs are running or the customer is present. In either case, the equipment is not producing a quality product, the printer loses customer confidence, and the printer is unable to generate cash flow. Overcoming equipment failures cannot be accomplished through quick-fix maintenance methods. The printer must develop a total production maintenance philosophy.

Preventive Maintenance Scheduling

It has been proven that a realistic, structured preventive maintenance program is one of the main elements in downtime reduction and optimizing equipment effectiveness. A battle over scheduling equipment maintenance continually rages between the manufacturing and sales/customer service areas within the printing organization. The question is, "Which is easier to deal with and control, scheduled equipment maintenance or unscheduled equipment failure?" If preventive maintenance does not become a serious part of production scheduling, then equipment breakdowns will become the production schedulers.

The printing organization must develop the discipline of adhering to established standards. The best way to develop the discipline is to implement the ISO 9000 or QS 9000 Quality Systems standards. Quality systems force the printer to document or say what they do, and then internal and third-party audits verify that they do what they say. Under subclause 4.9 (Process Control) of ISO 9000/QS 9000, the organization must maintain equipment to ensure process capabilities.

Under the QS 9000 standards a planned maintenance program must be established requiring documented procedures, consistent scheduling, replacement parts, and predictive maintenance management. Periodic preventive maintenance must be placed in the production schedule the same as a production job. The preventive maintenance results will include increased quality, increased capacity, and accelerated production throughput.

Summary

Elimination of production bottlenecks is necessary to minimize press downtime. Effective scheduling, preflighting files, using the highest-quality materials, and controlling variables on press will help to maximize your production capabilities by cutting chronic bottlenecks out of the process.

3 Statistical Process Control Tools

Identifying and eliminating causes of production stops and bottlenecks has already been recognized as the first step to accelerating production throughput.

The philosophy of continuous improvement is based on structured participation of people involved in the processes. True continuous improvement recognizes:

- Eliminating or killing the root causes to production problems will improve quality and productivity
- Work on the system, not around the system
- People want to do a good job
- The people most knowledgeable of the operation are the ones who do the job
- Everyone wants to be recognized for their valuable contributions
- All of us, the team, are smarter than one of us alone
- Structured team problem solving and SPC tools will accomplish more than individual efforts
- Recognize that everyone in the organization can be a major contributor to improvement
- An adversarial management/production staff relationship is outmoded and accomplishes very little

Problem-solving techniques for printing are a series of tools that the people within the organization use to solve problems and eliminate bottlenecks.

Flowcharts

To begin improving a process, the process steps must be identified and documented. Then each step in the process can be individually focused on, simplified, combined, and possibly eliminated. Flowcharts are a documented visual blueprint of the steps required to complete a process. The flowchart is a series of boxes and shapes connected by lines, visually

Sample flowcharts showing the make-ready process *(this page)* and the steps a job makes on the way to the press makeready process *(facing page)*.

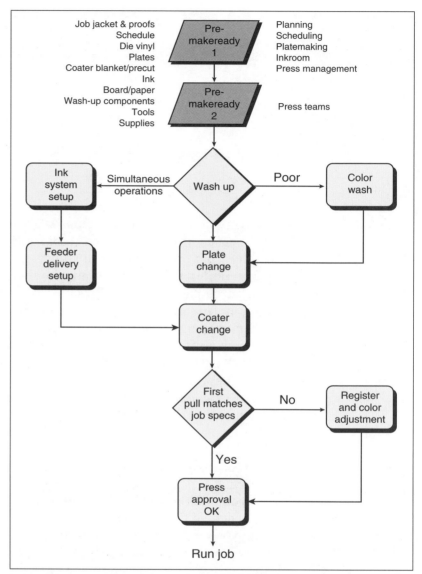

describing how a process is performed. The different boxes and shapes represent different steps and activities in the process.

One effective way to develop a flowchart is to form a team from the people who work in the process to be charted. In a brainstorming session, write down each step of the process on small pressure-sensitive note paper. Use different colored note paper to denote planning, input/output, process activi-ties, and decision points. Then the team can arrange the note papers on a wall in the sequence that the process actually flows. There are benefits from posting a flowchart:

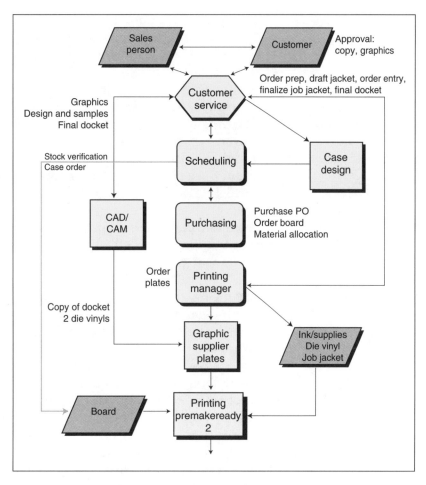

- The process flow can be documented in a visual format
- Procedures and work instructions can be effectively documented
- Flowcharts help teams to discover root causes of problems
- Flowcharts enable teams to restructure and improve the process

Standardizing the majority of the printer's operation activities is essential to minimizing problems and production stops. Developing flowcharts is the first step to developing and implementing a quality system, such as ISO 9000. Typical activities and processes include:
- Contract review
- Job estimate development
- Invoice processing
- Generation of purchase orders

- Contract proof approval
- Digital file preflight procedures
- Pre-makeready 1 and 2 activities
- Machine makeready
- Customer press approval
- Instrument calibration
- Handling of non-conforming products
- Preventive maintenance procedures and scheduling
- Handling of customer complaints
- Quality assurance of incoming materials
- Corrective action procedures

Documenting a process in a graphic format will enable the printer to quickly and effectively start the continuous improvement process.

Check Sheets   Check sheets are specifically designed lists usually used to track problem occurrences within a process. Check sheets can also be used as quality assurance and procedure tools to help prevent non-conforming products from occurring. Check sheets are designed to help track and measure the number of problems in a process. The check sheets help in pinpointing the root cause of those production problems and defects.

Quality-assurance check sheets may include plate inspection prior to going to the pressroom. Consistent and accurate plate inspection can minimize downtime in both the pressroom and platemaking areas. A sample plate-inspection checklist is shown on the following page, and in the Appendix.

Identifying and tracking printing defects is a more typical use of check sheets. Print-defect check sheets are used to identify any problems with the job prior to shipment to the customer. The data gathered from the check sheets must be used to identify and track occurrences of product defects. If the check sheet is used as a tool to criticize employees, its effectiveness as a process improvement tool will be lost. Check sheets can be customized for use in many areas of the printer's operation:

- Safety accidents
- Quality assessment of purchased product
- Production bottlenecks, types, and causes
- Digital preflight of files
- Digital file errors
- Plate remake reasons
- Paper waste and spoilage

- Production downtime in prepress, press, finishing
- Defective product
- Customer complaints
- Job reruns
- Errors in purchase product ordering
- Machine failure/breakdown
- Web breaks—occurrences and causes
- Excessive makeready time
- Excessive job completion time
- Machine idling and minor stops

Plate Inspection Checklist

❒ OK'd ❒ Blueline ❒ Double-check ❒ Matchprint present
         (do not check plate without this)
❒ Stripping checklist is present, signed, reviewed, OK
❒ Punch is correct (no punch damage)
❒ All burns are correct
❒ No spots
❒ No holes
❒ No halations
❒ No broken type
❒ Diamonds present for web
❒ Color bars included
❒ Trim marks included
❒ Register marks included
❒ Guide marks included
❒ Color breaks are correct
❒ Colors are labeled correctly and clearly marked for press
❒ PMS areas correct
❒ Varnish areas correct
❒ No varnish on gluetabs
❒ Roll off bars present
❒ Exposure is correct
❒ Content matches proof
❒ Color comparison for press present

❒ **OK to print**

Plate checked by_____ Date_____

Developing and using check sheets is among the best ways to gather data on the effectiveness of the printer's process. The gathered data can then be utilized in other SPC tools for effective process improvement.

Pareto Charts

Pareto charts are bar charts used to give a picture of the frequency of an occurrence within or resulting from a process. The higher the frequency of an occurrence, the higher the vertical bars are on the chart. Data from check sheets are entered into the Pareto chart, in descending order starting with highest frequency first. The criterion to be compared could be safety effectiveness, sales volume, profits, production costs, causes for plate remakes, machine downtime causes, percent of waste/spoilage, and reasons for late deliveries to customers.

The Pareto chart puts data into a visual form. That format greatly enhances communication for more effective process improvement. The Pareto chart is based on a process improvement concept known as the 80/20 rule. The 80/20 rule implies that 80% of the problem occurrences are the direct result of 20% of the potential production causes. The process improvement teams can immediately focus their efforts on eliminating 20% of the problems. The team can begin with correcting large problems and maximize their effectiveness over a short period of time.

Pareto chart showing the breakdown of a printing company's downtime.

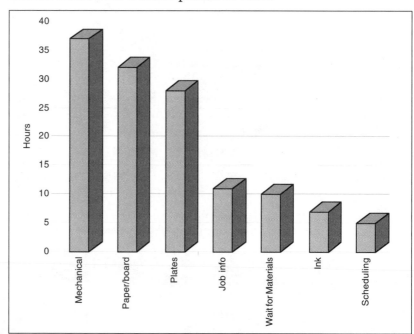

Run Charts

Run charts, also refered to as trend charts, are graphs that measure key process performances over specified time periods. The run chart reveals if the performance of a specified process is improving, declining, or stagnating. The control chart is usually used as a tool to determine if process improvement efforts have been effective. Run charts should have an historical performance baseline to graphically show where the process has been. The baseline indicator will be the reference point to determine the performance of the focused process improvement efforts. The measurement points on the chart can be documented and tracked weekly, monthly, quarterly, etc., depending on the indicator being measured.

Machine makeready and production per hour can be measured on run charts to determine if a process-improvement project on that machine has been effective. If the makeready improvement reveals times progressively lower than the baseline 90-minutes-per-makeready, then efforts are successful. If the charts reveal that print-sheet-production-per-hour increases above the baseline at the same time that makeready time decreases, accelerating throughput has been successful. Run charts can help track other key performance areas:

- Lost time on accidents per month
- Paper waste per job and per month
- Plate remake percentages per month
- Overall downtime hours weekly, monthly, and annually
- On-time delivery performance
- Machine failure/breakdown monthly, quarterly, annually
- Customer complaints
- Monthly sales volume
- Machine production per crewed hour per month
- Pre-makeready 1 completion rate per month
- Makeready time compared to an historical baseline over weekly and monthly time periods

It is essential that process improvement efforts be monitored for effectiveness. Run or trend charts are the tools that will determine if the program is going in the right direction.

Examples of run charts:

(Top) Major external factors cause spikes

(Middle) Shows inconsistent makereadies, possibly caused by experience levels of different crews

(Bottom) External factors are under control, lack of internal makeready focus is shown

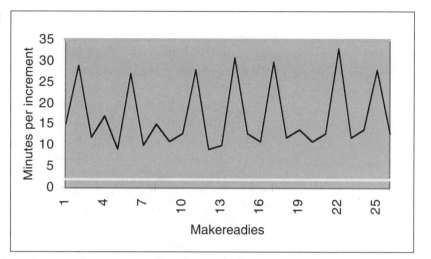

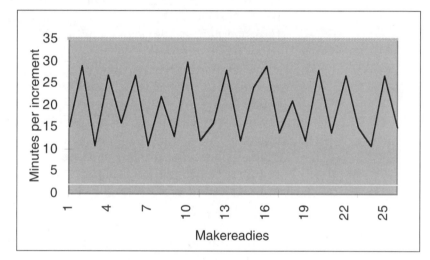

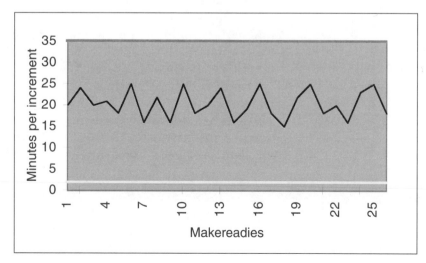

Control Charts

Control charts normally monitor product or process variability. Since all processes have certain amounts of variation, measured data will help determine if a process is under control. Control charts will have the recorded data points connected by lines. The chart usually has predefined target or average points, as well as upper and lower control limits. Control charts help to determine if products are good or bad, or if processes are in or out of control. Control charts can also reveal if changes in the process have occurred. Variability and attribute control charts can measure and monitor numerous process variables.

- Fountain solution pH and conductivity per shift
- Solid ink densities throughout a job to target ink densities
- Dot gains throughout a job to specified dot gains
- Makeready time compared to baseline for job
- Ink tack measurement
- Heatset web oven exit temperatures
- Number of production sheets run between blanket washes
- Paper/board loads moisture content
- Reported safety/accidents weekly

Control charts assist in gathering measurement data for analysis of process components to determine problems and if immediate corrective action is needed.

Sample control chart showing pH in relation to conductivity, over a period of hours.

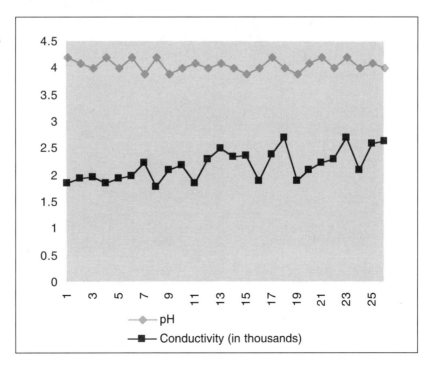

Histograms

Histograms are bar charts showing how consistent a process is to a central or target point or attribute. The more measurements recorded, the higher the bar cells will become. A histogram can take on different shapes based on the distribution of the measured data. A normal distribution will display a rather even bell-shaped curve with the highest bar cells located at and around the target point. Normal bell curves are the result of a consistently running process. A skewed bell curve, which can be either negative or positively skewed, is revealing that the distribution is consistent, but not as accurate to a target point. The bimodal curves show process consistency, centering around two different distribution points. One of the distribution points may be centered around the designated target point, the other could be either positive or negative to the target. A bimodal curve on a chart measuring the ink densities through a job could reveal one shift on the press ran the job to a different ink density point. Histograms can measure numerous attributes:

• Ink densities through a job and per press
• Dot gains through a job and per press
• Machine makeready times per press and job
• Caliper across new blankets
• Moisture content of printed loads
• Roller shore hardness from durometer measurement for new or older press rollers
• Platemaking exposure's solid-step points from targets

Sample histogram.

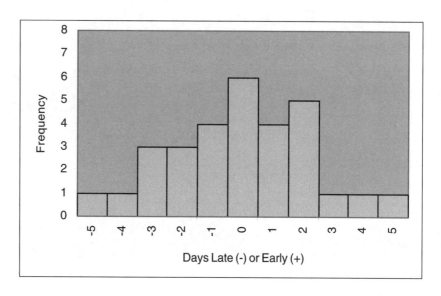

Scatter Diagrams   Scatter diagrams plot data from an attribute against data from a particular influence or variable. Scatter diagrams are designed to discover if changes occur from the variable's influence on the attribute being measured. If a variable increases, does the attribute improve or regress? One scatter diagram may show that press makeready times improve with more experienced press crews. Another scatter diagram may reveal lower dot gains when a sheetfed press is running faster than 10,000 iph. On the other hand, ink densities may become less consistent if a press runs above certain speeds, revealing a press or material problem. These measurement attributes are useful to chart:

• Makeready times versus press crew experience
• Paper size gain or loss (stability) versus relative humidity
• Press speeds optimum versus paper type or manufacturer
• Dot gain versus press speeds
• Ink density consistency versus press speeds
• Ink drying time versus different presses
• Plate resolution versus exposure time
• Coating gloss versus type of plate or blanket application
• Ink rub versus web oven temperature
• Plate remakes versus plate department shifts

Scatter diagrams will help the printer to determine what variables or causes influence changes in the process attributes. The printer can then take corrective and/or preventive actions to improve the process.

Sample scatter diagram showing the times for makeready in relation to the experience level of the press operator.

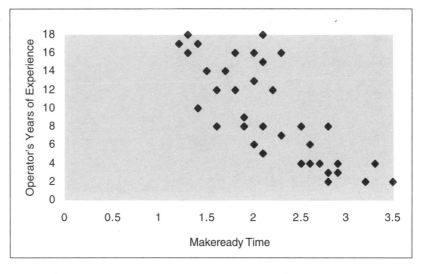

Cause-and-Effect Diagrams

Cause-and-effect diagrams, also referred to as fishbone or Ishikawa diagrams, are actually divided into two elements. The right side of the diagram normally states the effect or a problem element of a process. The left side of the diagram states the causes of a lack of improvement. Cause-and-effect diagrams are designed to illustrate and clearly organize causes that are affecting a process, or obstacles to improvement. There are usually a number of causes that prevent the optimization of a process. The causes are put into categories, such as manpower and people, machine and tooling, materials and supplies, methods and procedures, measurement, and plant and environment. Typically people who work in the process are formed into teams and meet in problem-solving or brainstorming sessions. The brainstorming session separately states a category and then lists the problems for that category. All the categories with related problems are then listed later on a fishbone diagram for solution development.

One of the major advantages of the fishbone diagram is that the lists form a visual picture of the process's problems. Using a fishbone diagram in the brainstorming meeting can be very effective when working with people who have no experience with team problem solving. From this visual list, the causes of and obstacles to improvement can be identified. The most striking causes that impact the process are then selected and analyzed for quick, effective corrective action. When analyzing selected causes, look for process-element changes, variations, and patterns in the main categories. One thing to remember when using fishbone diagrams is not to look for quick fixes or Band-Aids, look for root causes.

Cause-and-effect diagram, showing external elements contributing to quality makeready.

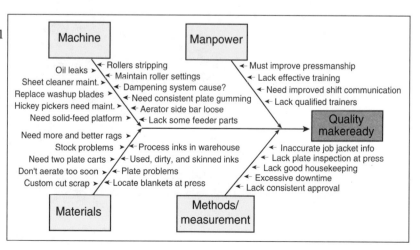

Processes and problems that cause-and-effect diagrams can help identify, organize, and improve are:
- Production or process bottlenecks (slow throughput)
- Inaccurate or missing job jacket information
- Causes for late deliveries
- Safety (back injuries)
- Machine idling and minor stops (time waste)
- Machine wait on materials downtime causes (time waste)
- Machine mechanical downtime causes (time waste)
- Causes for paper waste (material waste)
- Printed load ink blocking problems
- Hickey problems
- Obstacles to makeready improvement
- Causes for plate remakes
- Color variation complaints
- Reasons for job reruns
- Slow machine speeds

Cause-and-effect diagrams should be used as a visual continuous improvement tool. The cause-and-effect diagram can turn group problem solving from a finger-pointing exercise to an interesting and effective brainstorming session.

Summary

SPC tools are effective and necessary instruments for the printer to decrease waste. SPC tools are designed to work together in a step-by-step improvement process. The true effectiveness of SPC tools is how much the printer's team is involved and understands how the tools are used.

4 ISO 9000: An Operational Tool

The International Standards Organization (ISO) is comprised of nearly 100 member countries. By standardizing international practices, testing and certifications, the organization's goal is to prevent international economic problems. The ISO counterpart in the United States is the American National Standards Institute (ANSI).

ISO 9000–Quality Systems Standards provide structure and guidance for developing and implementing a structured quality system. ISO 9000 is a template that any manufacturing or service organization can plug processes, practices, and documentation into. A quality system is the set of activities, practices, and documentation that the printer must carry out to ensure quality-consistent products.

ISO 9000 Quality Systems Standards registration does not mean that every produced piece achieves both internal and external customer satisfaction. ISO 9000 registration means only that the implemented quality system is capable of satisfying customer requirements. One of the biggest benefits of being registered to ISO 9000 Quality System Standards is that the printer is forced to acquire discipline. Without discipline the printer will be unable to consistently adhere to operational procedures, preventive maintenance procedures, scheduling, and effective materials procurement.

The ISO 9000 series consists of five standards:

- **ISO 9000** provides guidelines and basic definitions. It also provides printers with guidance in selecting the appropriate standard to implement and certify to.
- **ISO 9001** is a model for printers to register their quality system to. It includes design and development, manufacturing production through handling and distribution, and delivery to the customer.

- **ISO 9002** is the same as 9001, except that it omits design/development activities and documentation. Because the design requirements are omitted, the majority of printers are registered to the ISO 9002 standard.
- **ISO 9003** is for organizations that require quality activities to ensure a quality-consistent product through final inspection and testing only. This standard omits activities and documentation for design/development and process manufacturing requirements. This standard is mainly for warehousing and distribution organizations.
- **ISO 9004** is a set of guidelines that a printer can use to assist in developing, documenting, and implementing a quality management system.

To achieve registration under the ISO 9000 standard, the printer must be audited by third-party certified auditors. Registrars are organizations that have been certified (in the USA, by the Registrar Accreditation Board [RAB]) to audit companies and award ISO 9000 registrations. ISO 9000 registrar auditors are trained and certified to perform audits of printer's quality system documentation and procedures, then report if everything examined is in conformance to the specific standard implemented. The auditors are not required to give assistance with any non-conformance discovered during the audit.

The following is a list of ISO 9000 clauses and sub-clauses, their documentation, and the reasons for implementing the standards. Also listed are typical problems that occur when a printer lacks a quality policy.

Clause 4.1

Management Responsibility

4.1.1	Quality Policy
4.1.2	Organization
4.1.2.1	Responsibility and Authority
4.1.2.2	Resources
4.1.2.3	Management Representative
4.1.3	Management Review
4.1.4	Business Plan
4.1.5	Analysis and Use of Company-Level Data
4.1.6	Customer Satisfaction

Why? Make certain that executive management are the visible leaders in defining, implementing, administering, and continually improving the quality system and meeting all customer requirements.

Management meets with a crew member to review the implementation of the quality policy.

What? The structured activities and documentation of executive management's responsibilities are delineated in clause 4.1. Executive management is responsible for establishing, reviewing, and providing the required resources to support the quality system.

How?
- Establish a quality policy
- Develop an implementation plan
- Define authority, responsibility, and assignments
- Identify required resources
- Appoint a management representative with executive authority
- Establish a management review process

When the printer is lacking:
- Top management is not truly committed or involved
- Quality policy statement is a marketing tool
- Customer complaints are more common
- Only short-term goals and little time to work
- Quality policy is implied, not defined
- Quality manager lacks defined authority
- No resources are designated to control and improve quality
- No designated review of quality system

Clause 4.2
Quality System

4.2.1 General
4.2.2 Quality System Procedures
4.2.3 Quality Planning

Why? Make certain that quality activities and practices provide graphic arts products that consistently meet customer satisfaction. Ensure continuous improvement of the quality system in the future.

What? Clause 4.2 shows, through documented procedures, the effectiveness of an operating quality plan or system, and that the system in operation supports the manual in reference to operation procedures and work instructions.

How?
- Determine the requirements of the ISO 9000 standard, including implementation and documentation
- Determine which ISO 9000 standard applies to the printer (9001, 9002, 9003)
- Plan the structure and outline of the quality system documentation: policy, manual, organizational chart, quality assurance organization, statement of authority and responsibility, distribution list of controlled copies, system clause 4.1–4.20, procedures index, forms index, operation procedures, work instructions, records
- Establish company practices with flow charts, procedures, and work instructions
- Evaluate current and needed resources
- Establish planning functions to meet requirements of the ISO 9000 standard
- Implement the quality system by being aware of quality plans, resource and time requirements, continual updates of procedures and documentation, measurement requirements and latitudes, acceptance standards

When the printer is lacking:
- Quality system and manual are non-existent
- Lack documented procedures
- Documented quality procedures are not consistent throughout the organization

Clause 4.3
Contract Review

4.3.1   General
4.3.2   Review
4.3.3   Amendment to a Contract
4.3.4   Records

Why? Ensure that the printer is able to meet the customer's requirements before accepting the job.

What? Documentation ensures that all the customer's requirements are adequately defined, and that the printer understands them. The printer must follow procedures for reviewing requests for quotations, contracts, or accepted job orders for both new and rerun jobs. All differences must be resolved.

How?
- Document customer requirements
- Identify precontract practices
- Establish contract review procedures
- Verify the printer's capabilities to meet job requirements
- Internalize customer requirements; resolve any differences
- Maintain control of customer purchase orders for single- or multi-contract orders
- Develop deployment plan
- Establish customer purchase order review procedures
- Obtain customer agreements
- Track, improve, and revise contract review procedures

When the printer is lacking:
- Lack thorough internal reviews of job requirements before submission or acceptance of jobs
- Lack knowledge of printer's capabilities to print new jobs
- Customer planning lacks clear, consistent procedures

Clause 4.4

Design Control (ISO 9001 only)

4.4.1   General
4.4.2   Design and Development Planning
4.4.3   Organizational and Technical Interfaces
4.4.4   Design Input
4.4.5   Design Output
4.4.6   Design Review
4.4.7   Design Verification
4.4.8   Design Validation
4.4.9   Design Changes

Why? Ensure that the product meets all design requirements specified by the customer and any regulatory agencies.

What? Clause 4.4 details procedures to control, verify, and validate product design and support software.

How?
- Document all customer requirements
- Establish design control planning; assign duties and responsibilities
- Assign quality design staff and provide adequate resources
- Obtain input from all departments
- Document control procedures
- Design output to meet input requirements and relevant regulations; include patent reference data; review before release
- Provide output verification
- Validate the design
- Develop or change control procedures

Clause 4.5

Document Control and Data Storage

4.5.1   General
4.5.2   Document and Data Approval and Issue
4.5.3   Document and Data Changes

Why? Ensure that all quality system documents and data are under control and easily accessible to all people when they need them.

What? Clause 4.5 lists methods and procedures to ensure that documents and data are accessible, updated, reviewed and revised periodically, and controlled in hard copy and electronic digital form.

How?
- List all documents
- Establish plans to administer each category of documents
- Ensure accessibility of documents at all work sites
- Establish control of obsolete documents
- Establish document-change control procedures
- Investigate and review conformity to change procedures

When the printer is lacking:
- No document control exists
- Documents and procedures are issued as bulletins posted on boards around the organization
- No one in the organization knows which documents are valid and which are not

- No procedures exist to review documents in the system
- Obsolete documents still float around

Clause 4.6
Purchasing

4.6.1   General, Approved Materials for Ongoing Production
4.6.2   Evaluation of Sub-Contractors
4.6.3   Purchasing Data
4.6.4   Verification of Purchased Product
4.6.4.1 Supplier Verification at Sub-Contractors' Premises
4.6.4.2 Customer Verification of Sub-Contracted Product

Why? Ensure that products received from sub-contractors meets the printer's requirements. Products include hardware, software, process materials, and services.

What? Clause 4.6 lists methods and procedures to ensure effective purchasing of required process materials, hardware, and software, including sub-contractor evaluation, purchasing data, and verification of purchased products and services.

How?
- Evaluate current purchasing specifications and requirements
- Upgrade specifications as needed
- Prepare, review, and approve purchasing documents
- Establish criteria for sub-contractor acceptability
- Develop sub-contractor classification system
- Establish a realistic record-keeping system

When the printer is lacking:
- Materials are purchased on price alone, the lowest bidder wins despite merit
- No evaluation exists to determine acceptability of manufacturers and suppliers
- Visits to suppliers are simply tours plus lunch
- Printer does not provide suppliers with adequate materials specifications
- Verification of incoming materials is lacking

Clause 4.7
Control of Customer-Supplied Product

Why? Ensure that the product received from the customer is accurate and compatible with the printer's process. The customer-supplied product must ultimately meet all customer requirements for the final product.

What? Clause 4.7 documents procedures for verification, storage, and maintenance of customer-supplied products.

How?
- Determine the existence of the customer-supplied product
- Document the current practice for verification, storage, and maintenance
- Revise and improve the printer's procedures
- Evaluate revisions and effectiveness

When the printer is lacking:
- Customer-supplied films are not proofed to verify accuracy to provided proofs
- No preflight is done to customer-supplied digital disks
- No verification of customer-supplied brokered paper

Clause 4.8

Product Identification and Traceability

Why? Ensure that the printer's product is properly identified at all stages of production, and prevent errors, scrap, and rework.

What? Clause 4.8 lists procedures for identifying incoming materials, work-in-process, and finished products. Identification is the ability to separate two or more materials or products. Traceability is the ability to separate materials or products by individual unit batch or run.

How?
- Establish customer or regulatory requirements
- Document current traceability practices to include information from the subcontractor, in the printer's plant, and to the customer
- Review and revise traceability procedures
- Determine types of traceability/identification—units, lots, production dates
- Determine methods of identification—hardcopy, digital, labels, bar codes
- Determine marking medium—manual permanent marker, colors, computer printer
- Record keeping—availability, retention time, responsibility

When the printer is lacking:
- Last load of stock is found after the job is pulled off press
- Stock is not found for job
- Wrong stock is brought to press
- Job is delayed because a new shipment of plates could not be found

Clause 4.9

Process Control

* Government Safety and Environmental Regulations
* Designation of Special Characteristics
* Preventive Maintenance
4.9.1   Process Monitoring and Operator Instructions
4.9.2   Preliminary Process Capability Requirements
4.9.3   Ongoing Process Performance Requirements
4.9.4   Modified Preliminary or Ongoing Capability
         Requirements
4.9.5   Verification of Job Setups
4.9.6   Process Changes
4.9.7   Appearance Items

Why? Ensure that all processes are carried out under controlled conditions to minimize manufacturing variability.

What? Clause 4.9 establishes standards for documenting procedures; identifying and planning production steps; ensuring that all equipment is suitable and operating under controlled conditions; ensuring a suitable working environment; preparing documented instructions for all activities affecting quality; approving and monitoring processes; establishing workmanship criteria where required; and maintaining equipment to ensure process capabilities.

How?
• Base process control on the quality plan
• Identify critical control points
• Define factors affecting key process controls: equipment, environment, and hazardous materials control

- Identify product requirements: specifications, workmanship, and regulatory standards
- Review current monitoring techniques
- Develop control and approval procedures
- Develop work instructions
- Develop equipment maintenance procedures
- Identify special processes
- Implement process change controls
- Revise and improve procedures
- Evaluate revisions

When the printer is lacking:
- Preventive maintenance is lacking or poorly done
- No testing and calibration of equipment is done
- Housekeeping and environmental controls are poor
- Documented procedures and work instructions are lacking or not followed
- Specified proofs, samples, and standards match poorly at press

Clause 4.10

Inspection and Testing

4.10.1	General
	* Acceptance criteria
	* Accredited laboratories
4.10.2	Receiving Inspection and Testing
4.10.3	In-Process Inspection and Testing
4.10.4	Final Inspection and Testing
4.10.5	Inspection and Test Records

Why? Ensure that all products conform to requirements at every step of the production process; identify nonconforming products, at the earliest possible stages, and be able to facilitate corrective actions.

What? Clause 4.10 lists procedures for inspection and testing throughout process operations to ensure:
- Incoming product
  - Procedures for inspection and verification
  - Controlling until verified
  - The amounts of inspection and testing depends on subcontractor control
- In-process product
  - Procedures for identification and inspection of product
  - Monitoring processes
  - Product control until tests are completed

Testing with a scanning densitometer.

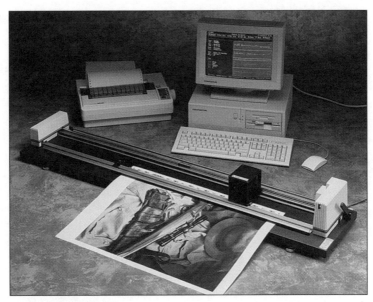

- Finished product
  - Procedures ensuring that all inspections and tests are completed
  - Product conforms to requirements
  - Product is not released until inspection/testing is complete

How?

- Establish separate plans and procedures for receiving inspection/testing, in-process inspection/testing, final inspection/testing
- Determine policies
- Identify categories of all products that are affected
- List all quality characteristics that are subject to inspection/testing
- Make sure that procedures for identifying specified requirements are easily accessible
- Provide complete and current procedures at points of inspection
- Provide positive product identification/recall for urgent release
- Release product only when tests and records are complete
- Revise and improve procedures
- Evaluate revision

When the printer is lacking:

- Paper/board quality is not tested when delivered
- Ink proofs are not requested by printer

- Incoming films or discs are not inspected or proofed
- Press sheet inspection frequency is not consistent through pressrun
- Ink density control on press is inconsistent, some press operators use densitometers other use visual inspection
- Ink densities are not documented

Clause 4.11

Control of Inspection, Measurement, and Test Equipment

4.11.1   General
4.11.2   Control Procedures
4.11.3   Inspection, Measuring, and Test Equipment Records
4.11.4   Measurement Systems Analysis

Why? To ensure that inspection, measuring, and testing equipment are capable of consistently providing specified data requirements, so proper control and acceptance decisions can be made.

What? Documented procedures to ensure that equipment is properly calibrated and maintained. Also that measurement uncertainties are known and are consistent with required capabilities.

How?
- Identify all inspection and test requirements
- List all available equipment and software
- Identify calibration requirements and verification procedures for all test equipment
- Plot a flow chart and review current procedures and documentation for:
  - Measurements to be made
  - Calibration procedures
  - Identification of calibration status on the equipment
  - Corrective actions
  - Working environment control
  - Handling and storage
  - Safeguards against unauthorized adjustments
  - Rechecking/calibration intervals
- Revise and improve procedures
- Determine hard copy versus digital records
- Establish effective record keeping
- Evaluate revisions

When the printer is lacking:
- Densitometers are not calibrated to a T-Ref
- Frequency of densitometer calibration is unknown
- Conductivity and pH meter calibrations are unknown
- Durometer calibration is not checked
- Packing gauge calibration is not checked
- Viewing booth lighting is not consistent (5000 K)
- Spectrophotometer calibration is unknown

Clause 4.12
Inspection and Test Status

* Product Location Identification
* Supplemental Verification

Why? Ensure that only the product passing through the required inspection and testing points are released.

What? Clause 4.12 establishes procedures for the identification of inspection and test status throughout the production process: labels, tags, stamps, markers, status cards, inspection records, software programs, and production locations.

How?
- Identify locations where inspection is critical: receiving, pre-production, process production, post-production shipping
- Chart all processes
- Determine the identification medium: marking pens, colors, stamps, tags, labels, staging areas, hard copy, and digital records
- Review positive release procedures and responsibilities
- Revise and improve quality plan and procedures
- Evaluate revisions

When the printer is lacking:
- Printed load status is not identified
- Non-inspected printed loads are released to the next stage
- Plate inspection and identification are not marked on plate

Clause 4.13
Control of Non-Conforming Product

4.13.1 General
4.13.2 Review and Disposition
4.13.3 Control of Reworked Product
4.13.4 Engineering/Quality Assurance Approved Product Authorization

Why? Ensure that the printer does not use or deliver non-conforming products to internal or external customers.

Non-conforming product; in this example, blistering is shown.

What? Clause 4.13 establishes operational procedures that will identify non-conforming products; evaluate the extent of non-conformity; segregate the non-conforming product from the conforming product; define authority for disposition of non-conforming product; and notify all parties concerned.

How?
- Review and document procedures for identification, documentation, segregation, and prevention of mixing non-conforming products with conforming products when delivering to the customer.
- Document procedures for disposition, notification, and classification
- Assign authority of disposition approval
- Document procedures for reinspection of reworked jobs
- Document reporting and handling procedures
- Revise and improve the procedures
- Evaluate revisions

When the printer is lacking:
- Printed loads with quality problems are not identified
- Print/quality problems are not marked with color codes
- Print/quality problems are overmarked or overstated
- Print/quality loads are poorly segregated
- No stage location exists for loads with print/quality problems

Clause 4.14

Preventive and Corrective Actions

4.14.1   General—Problem Solving Methods
4.14.2   Corrective Action
4.14.3   Preventive Action

Why? Ensure proper investigation of non-conforming products, and institute ways to eliminate them. The printer is to attempt detection and elimination of potential causes of non-conforming products before the problems occur. This is the main continuous improvement function of ISO 9000.

What? Clause 4.14 lists how to investigate and eliminate causes of non-conforming products at all points in the production process, including handling and distribution to the customer. This clause works to prevent the occurrence of non-conforming products.

How?
- Separately identify corrective action procedures from preventive action procedures
- Carry out corrective actions
  - Assign responsibilities
  - Review the type and number of complaints
  - Evaluate impact on operational costs
  - Evaluate effectiveness of the current action
  - Provide resources
  - Revise and improve procedures
  - Institute procedural changes
  - Evaluate the revised procedure's effectiveness
- Carry out preventive actions
  - Assign responsibilities
  - Review existing preventive action procedures if any
  - Evaluate current practices
  - Identify or establish new preventive action activities
  - Improve procedure effectiveness
  - Report preventive actions when taken
  - Review and evaluate improved procedures
  - Submit recommended actions to management review

When the printer is lacking:
- Quick-fix problems, there is no focus on root causes
- Managers solve problems all the time, fire fighting
- No corrective action strategy exists
- No time to take preventive action

- "That's close enough" attitude is prevalent—i.e., the press operator can fix it on the press
- Same problem occurs each time that job runs
- Downtime on the press is downtime of the entire process, or temporary bottlenecks

Clause 4.15

Handling, Packaging, Storage, and Delivery

4.15.1	General
4.15.2	Handling
4.15.3	Storage
	* Inventory
4.15.4	Packaging
	* Customer Packaging Standards
	* Labeling
4.15.5	Preservation
4.15.6	Delivery
	* Supplier Delivery Performance Monitoring
	* Production Scheduling
	* Shipment Notification

Why? Ensure that procedures for handling, packaging, preservation, and delivery of the printer's product are adequate to protect product integrity at all stages of production, storage, and delivery.

What? Clause 4.15 establishes standards for documenting procedures in order to ensure product integrity for all stages in the process:
- Incoming materials
- Work-in-process product
- Finishing product
- Specified delivery conditions
- Deterioration and damage prevention
- Secure storage and disposition procedures
- Packing and packaging methods
- Product segregation

How?
- Identify critical points in the process
- Review information concerning shelf life and damage type and frequency
- Develop documentation for:
  - Process handling procedures
  - Warehouse safety procedures
  - Package designs

○ Inventory management procedures
○ Customer-specified segregation requirements
○ Storage/preservation/segregation methods
○ Transportation carrier selection
○ Environmental concerns
• Develop preventive maintenance of materials handling and packaging equipment

When the printer is lacking:
• Skids of paper stock tip over during transport to press due to lack of personnel training
• A paper roll is dropped and damaged because the roll clamp not working right
• The printed job is damaged in carrier truck because a weak skid broke
• Inadequate corrugated is used to ship job because the customer shipping requirements were not reviewed
• Leak in the warehouse roof damaged a job stored in corrugated cartons

Clause 4.16

Control of Quality Records

Why? Ensure that the printer's quality records are effective and clearly show the quality-system operation.

What? Clause 4.16 lists the documented procedures effectively giving control of quality records. Effective quality records control includes adequate records storage, established record retention time, accessibility of quality records for customer review, and quality records legibility.

How?

- List and review documents
- Review procedures for file collection, identification, access, filing, storage, maintaining, and disposition
- Review quality records of suppliers, manufacturers, and subcontractors
- Establish document integrity criteria for legibility, identification, retention, storage, and customer access
- Review quality records requirements under the pertinent ISO 9000 clauses: 4.1.3, 4.2, 4.3.4, 4.4, 4.6.2, 4.7, 4.8, 4.9, 4.10, 4.11, 4.12, 4.13, 4.14, 4.17, and 4.18

When the printer is lacking:

- No quality documentation is kept
- A collection of memos and directives are scattered all over the building
- Not everyone knows where the quality records are
- Quality records are not complete, numerous documents are lost
- Customers visits turn into dog and pony shows

Clause 4.17

Internal Quality Audits

Why? Ensure that the printer's quality system meets requirements and helps the printer to objectively monitor themselves. Internal auditing ensures both the effective maintenance of the quality system, and the continuous improvement of their process.

What? Procedures for planning and conducting internal audits of the quality system. Activities include training personnel to be internal auditors, scheduling audits, conducting the audits, documenting audit results, communicating audit results during management review, and verifying effectiveness of corrective actions taken.

How?

- Identify areas to be audited
- Establish auditor qualifications (experience, attitude, and training)
- Develop audit procedures, planning, and documentation
- Conduct initial quality audits; evaluate effectiveness of procedures, verify compliance to documentation
- Establish an ongoing quality audit program

Perform a quality control audit, documenting any findings during the process.

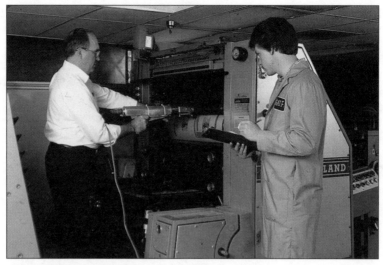

When the printer is lacking:
• If a quality system has been developed, no one knows if it is working
• The quality system is scrapped every time things get busy

Clause 4.18
Training

Why? Ensure that employees are properly trained to do their jobs and be able to effectively address new and future technology. Training will better enable employees to maintain quality and avoid mistakes.

What? Clause 4.18 works to develop training that encompasses both quality system requirements and process and technology requirements. Requirements must include analyzing training needs, developing a training plan, training personnel based on those requirements, and keeping individual records.

How?
• Identify training needs, listing knowledge requirements and job skills
• Develop and document a basic training plan
• Provide training based on the quality system, knowledge of equipment, and workmanship requirements

- Establish and record all training in personal files

When the printer is lacking:
- No training exists, knowledge is assumed to be inbred
- No review is conducted or strategy developed for organizational training needs
- Training program is not structured with purpose or goals
- No way exists of verifying if training is actually effective

Clause 4.19
Servicing

Why? Make sure that after-sales servicing is provided if required, to help ensure complete customer satisfaction.

What? Clause 4.19 calls for documented procedures to maintain and verify if servicing requirements are being met.

How?
- Identify customer service requirements
- Document service requirements
- Revise and improve procedures
- Evaluate effectiveness of revisions

When the printer is lacking:
- Few printers are required to perform defined service for their product after delivery to the customer.

Clause 4.20
Statistical Techniques

4.20.1   Identification of Statistical Requirements
4.20.2   Procedures
    * Selection of statistical tools
    * Knowledge of basic statistical concepts

Why? Ensure that the printer has identified steps in the process where the installation of statistical measurement techniques maintain quality and detect possible quality problems.

What? Clause 4.20 requires the identification and application of effective statistical measurement tools and techniques for verifying and controlling process capabilities and product quality. The clause establishes documented procedures to ensure that statistical techniques are correctly and consistently applied.

How?
- Identify current statistical techniques and procedures

- Review the effectiveness of statistical techniques
- Determine if additional applications are needed
- Establish training needs
- Select personnel to be trained for working in SPC areas
- Conduct the training program with expert assistance
- Review effectiveness of new or expanded SPC applications

When the printer is lacking:
- No SPC implementation exists
- SPC charts are posted in the break room or given to middle managers as a "gotcha" tools
- SPC data is input on charts and posted, nothing is done with the data

Summary

ISO 9000 has clauses or elements that are designed to continually improve the printer's process. They include:

4.1.1	Objectives of Quality
4.1.3	Management Review
4.4	Design Control
4.14	Corrective and Preventive Action
4.17	Internal Audits
4.18	Training
4.20	Statistical Techniques

ISO 9000, if implemented based on the true intent, is a very effective process improvement tool. ISO 9000 will provide the management with discipline that many printing managers lack for eliminating downtime and establishing a very effective total production maintenance system.

5 Quality Assurance of Print Materials

The printing process has become extremely complex. Technology has advanced to the point where printing, once considered an art form, has evolved into a science. In the past, where variations and defects in printed products had been accepted as merely part of a process, customers are now demanding higher and more consistent quality than ever before, placing demands on printers to focus their attention on quality controls.

Admittedly, printers have little control over the manufacturing quality of pressroom materials such as paper, blankets, inks, plates, and rollers. And most printers do not have the time or financial resources to conduct extensive in-house quality testing on pressroom materials; they basically depend on manufacturers and suppliers to provide quality products. Some commonsense strategies, however, can help printers minimize problems before a job is on press.

Quality Standards: ISO 9000

One way to institute policies and procedures for quality control is by implementing the ISO 9000 Quality System Standards. Three clauses of the standard deal directly with consistent control of incoming materials. The earlier that potential problems are caught, the more efficient the process becomes. ISO 9002 Quality Standards (ANSI/ASQC Q9002-1994) are the ones most pertinent to the printing industry.

For more information or to order your copy of the ANSI/ASQC Q9002-1994 standards, contact the American Society for Quality Control (ASQC) at 1-800-248-1946.

Establishing Standards, Procedures

An excellent starting point is to establish standards for selecting and purchasing materials before they are inevitably involved in the pressroom quality control process. Vendors should be evaluated and ranked on the basis of performance, implementation of quality programs, customer relations, willingness to participate in customer quality improvement efforts, and costs. Selection based on price alone is not recommended. Although selecting a single supplier for each type of material is ideal, reliable secondary suppliers should be considered. Clear, concise, and realistic expectations and specifications should be established and documented. Seeking a supplier's assistance in devising these expectations can help to ensure clear communication.

Purchasing

When placing orders with suppliers, it is extremely important to provide complete and accurate specifications. Failure to do so can result in loss of time and money, and you may receive products you didn't order. The best time for standardizing and establishing specifications for each type of product you order is well in advance of placing the order.

Inspecting Incoming Materials

Each incoming material should be inspected following a set procedure or checklist. At the point of receipt, make sure you receive what you ordered. Verify the material's purchase order, examining the material's type, age, size, and quantity for overall quality and accuracy. Look for external damage. If damage has occurred, it is often visible and should be documented and reported immediately. More extensive inspection can be conducted in the pressroom, but examining the materials at the time of delivery saves time.

Handling and Storage

Materials received in acceptable condition can often become corrupted when stored improperly. Proper ambient conditions—temperature, relative humidity, and lighting, for example—must be maintained, particularly for storing such materials as paper, plates, and chemicals. Steps should also be taken to ensure proper handling of containers. Employees using forklifts, for example, should be properly trained.

Inventory

Maintain accurate records so that information can be readily retrieved on shipment quantity, shipment dates, warehouse location, etc. Establish procedures for placing materials into the inventory so that inventory levels can be monitored daily. It is recommended that inventory be based on a revolving "first-in, first-out system" to prevent dated materials from accumulating or being inadvertently used. Materials rejected for any reason should be removed from the system entirely to prevent miscalculations of stock levels.

Record Keeping   Maintaining accurate records about the performance and conditions of materials is especially important if nonconforming materials must be replaced or reimbursed. The records will also help the supplier locate the origin of the nonconforming product and take corrective and preventive actions. Suppliers may also be helpful in initially determining what performance factors are pertinent.

Standardizing policies and procedures for the basics—purchasing, inspecting incoming materials, handling and storage, inventory, and record keeping—will help you markedly improve the quality of pressroom materials. These fundamental steps can save time, money, and frustration for all parties—printer, supplier, and customer.

Paper and Board

Paper/board producers use elaborate testing equipment and routine test procedures to ensure that the end product does not deviate from control limits. Tests such as caliper, picking, absorption, tear and burst strength, brightness, and color enable them to predict longevity and performance. Most printers do not have the resources to conduct these tests but can still develop procedures to minimize difficulties stemming from paper. Even the paper manufacturers still rely on the printing press as the final determination of printability.

Verify the quantity and specifications of shipments on receipt.

Specifications   Detailed, well-thought-out specifications will greatly improve your control over the quality of incoming shipments. Develop a checklist using the suggestions listed here. Your own list may need more or less detail but can include the following:

- Brand name
- Color
- Type of paper (coated two sides enamel [C2S], coated one side label [C1S], bond, newsprint, solid unbleached sulfite [SUS], solid bleached sulfite [SBS], clay coated newsback [CCNB])
- Basic size (for example, 25×38 in.)
- Basic weight (for example, 70 lb.)
- Caliper (for example, 0.018 in.)
- Grain direction (long or short)
- Sheet size and quantity sheet count. Indicate whether the paper needs to be supplied in flat sheets trimmed on all sides for work-and-tumble and perfecting jobs.
- Packaging requirements
  ○ **Skids.** Specify single-stacked or double-stacked and maximum skid height. Ask for skids suited to your operations. For example, will the paper be off stacked or go directly into the press feeder?
  ○ **Cartons.** Specify maximum sheets per carton and carton dimensions, whether stacked on skids or pallets, double-stacked or single-stacked, and list the maximum stacked height of the cartons.
  ○ **Rolls.** Specify the amount in weight (pounds), specified web width, maximum roll diameter, core specifications (inside core diameter), core type (returnable or nonreturnable), and maximum splices per roll. Indicate the maximum allowable variation in roll width for web presses or sheeting operations.
- List special labeling instructions or information (e.g., roll position from mill) to be noted on protective outside wrapper on roll stock.
- Note shipping/delivery instructions (delivery date and location, type of vehicle for shipment such as truck or rail, and loading instructions for transport—rolls on side, skids, pallets, maximum stacking requirements).

Supplying the paper vendor with all pertinent information at the time of the order will help prevent problems in the printing process and can also save valuable time. If you receive the wrong paper because your ordering information was incomplete and if your order was a mill order rather than a stock order, it may take weeks to re-ship your paper order.

Inspection

As with all pressroom materials, thoroughly inspect paper at the time of receipt. A checklist can be helpful for documenting the paper conditions at the time of delivery. Check the packing slip to confirm that the shipment is correct.

Visually inspect the shipment before unloading the truck or railroad car. During unloading, look at each skid or roll for external damage, such as torn or gouged wrappers, crushed rolls (remove a section of wrapper to check for a crushed core on suspect rolls), or collapsed skids or pallets. Take a picture of any damaged roll or skid and document the type of damage on the checklist. Have the carrier's representative acknowledge the condition of the shipment and sign off on both good and bad conditions before the load is accepted.

Verify the quantity (sheet count for skids and cartons, weight for rolls) and specifications that should be on the wrapping labels (brand, size, type, basis weight, caliper, color). Indicate any deviations on the checklist.

You can cut through the wrapper of a roll or skid to remove small samples of the paper for inspection. Remember to reseal the cut-out hole in the wrapper immediately with the piece of wrapper and shipping tape to avoid moisture or other contamination. The samples must be put in plastic bags right away and sealed for protection. Tests conducted by a third party may be necessary for any problems incurred. Caution: consult with the paper manufacturer before cutting into any wrapper. Some manufacturers feel that the integrity of the moisture barrier is affected when the wrapper is removed. If this is the case, you can request mill out-turns for inspection prior to acceptance of a shipment.

Damage to the wrappers on rolls of paper can result in loss of necessary moisture, or in too much moisture.

Request run control and test data from the paper manufacturer for each mill run. The printer can conduct tests on the paper. It is recommended that paper manufacturers be consulted concerning procedures and expected results.

1. Check moisture content in skid loads of paper with a sword hygrometer
2. Use Dennison pick test waxes on paper or clay-coated board to determine potential picking problems
3. Conduct absorption tests using K&N or Linnetta test inks to determine ink hold-out and potential paper mottle
4. Use a colorimeter or spectrophotometer to determine color and brightness

Handling and Storage

Proper handling and storage of paper ensures that it will perform to expectations, and it can minimize your liability if problems do occur:

- Make sure storage areas have adequate space and are clean
- Establish maximum stacking spacing specifications for both rolls and skids of paper
- Immediately document any damage that occurs while placing the paper into storage
- When stock is removed from inventory and taken to the pressroom, visually inspect for damage and document the results
- Move stock to the press staging area a few days before the pressrun so it can acclimate to pressroom environment

Proper storage for paper, raised on a skid with a plastic cover.

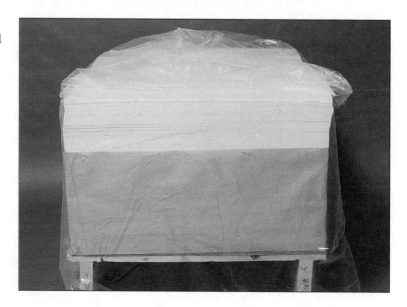

- If rolls or skids do not show any visual damage at the time of receipt, but paper is later discovered to have damage, take pictures, save samples, and document the skid and roll numbers. This information can be used when consulting the paper supplier concerning the problem.
- Cover any open skids and partial rolls of paper with plastic (preferably plastic stretch wrapping) to protect them from dirt, dust, and environmental changes.
- Mark skids or rolls with easily identifiable load tags.
- Establish stacking specifications for partially used skids and rolls to prevent damage during storage.
- Keep power lift trucks and roll clamps in proper working order to prevent damage to rolls or skids.
- Train employees to properly use the established inventory system, lift trucks, roll clamps, and other equipment.

Plates

Ordering and storing plates can be a complex issue. Again, the appropriate policies and procedures can minimize problems due to light, temperature, and humidity sensitivity, and to the resulting pressroom complications.

Specifications

The type of plate or process type (lithography or waterless, medium or long run) you order depends on a job's quality and run-length requirements. A purchase order for plates should include the following:
- Brand name
- Type (negative or positive and aqueous or solvent)
- Dimensions
- Caliper thickness
- Dimensional tolerance (particularly for web presses)
- Quantity
- Shipping and packaging instructions, including number of plates per corrugated carton, maximum number of plates per crate, type of slip sheets between each plate, moisture barrier requirements, and transport means (freight or overnight)

Inspection

When the plates are delivered, compare the invoice, shipping order, quantity, type, and size of the plates with the information on the purchase order to ensure accuracy. Again, shipping containers should be thoroughly inspected for damage. If damage is discovered, photograph it and contact both the supplier and the carrier immediately. The decision to accept or reject the shipment should be based on the extent of the damage. Remember, plates are extremely sensitive to environmental conditions and light.

The platemaking department should conduct routine exposure tests to determine if the plates, exposure equipment, and processor are conforming to specified requirements. Using a sensitivity guide such as the GATF Plate Control Target (see chapter 8), conduct exposure tests on each new shipment of plates and at least twice a week to determine if the resolution is acceptable or the exposure is correct. On the vacuum frame, expose the targets in the center and four corners to determine if illumination of the vacuum frame is even or if the light source falls off on the outer edges. Determine exposure consistency on a step-and-repeat machine by exposing up to eight steps across the plate. After exposure testing, compare the results of the tests to manufacturer specifications for exposure precision.

Handling and Storage

Shipping cartons should remain sealed until they are delivered to the platemaking area. Unexposed plates should be taken out of their cartons and handled under yellow safelight conditions to prevent potential fogging before processing. Specify the maximum number of cartons that can be stacked and use plates on a first-in, first-out basis so old or outdated plates do not accumulate in inventory. The plate storage area should be maintained at 70–75°F, relative humidity of approximately 45–55% RH. This will minimize static electricity and help maintain film dimensional stability.

Plate processors should be cleaned and chemicals changed regularly, based on manufacturer specifications. Notify the manufacturer immediately if problems with processing develop after preventive maintenance.

Keep records of routine exposure test results and of scheduled and unscheduled maintenance on exposure units, plate punch, and processor. Inspect plates for spots, voids, mistakes, broken type, correct color, halation in screens, quality of halftones, and position. Document plate remakes and chart them to determine the origin of a problem.

Ink

Poor ink quality is detrimental in printing. Not every printing plant has an on-staff chemist to conduct testing so again, printers must rely on manufacturer recommendations and test results to choose the best inks. Your ink supplier can assist you in testing inks. This practice is highly recommended to ensure that the type of ink you choose suits your

requirements. An ink
supplier can provide the
following information
with each order:

- Tack rating, including
  number and revolu-
  tions per minute
- Viscosity rating
- Print from a proof
  press on the stock the job will be run on press, at the
  accepted ink film thickness for that type of ink (0.2–0.4
  mil normally for process sheetfed inks)
- Spectrophotometer data

Other data gathered from the ink test runs such as density, dot gain, mileage, drying characteristics, gloss, rub
resistance, ink lay on the sheet, ink/fountain solution compatibility, and price will enable the printer to make sound,
qualified decisions.

Specifications   Again, specifications to the supplier should be clear and
concise. Specifications to ink vendors should include the
following:
- Quantity
- Color sample swatch
- Formula number
- Type of containers (1-lb. cans, 25-lb. kits, etc.)
- Delivery date required
- Delivery method
- Color sequence
- Type of printing process used
- Type of substrate the ink will be printed on (uncoated
  paper, coated paper, clay-coated board, polycoated board,
  plastic)
- Type of fountain solution
- Drying requirements (conventional or quickset)
- Coating type, if any (resin varnish, aqueous, UV, and EB)
- Press speed
- Rub resistance
- Fade resistance
- Resistance to specific chemicals

The more information your supplier has about the job specifications, the more your supplier will be able to recommend appropriate inks.

Inspection

Inspection of inks is limited to checking for quantity, visible damage, and accuracy in comparison to the purchase order. Laboratory testing is possible, but often not feasible. The following recommendations will reduce ink problems:
- Use a first-in, first-out inventory system
- Firmly reseal all partially used containers to avoid drying
- Call the ink manufacturer promptly if problems develop
- Verify compatibility of inks for different substrates
- Avoid ink additives for adjusting inks on press
- Consult with your ink vendor to ensure overall quality and compatibility

Matching PANTONE® and special colors accurately can be difficult and in some cases almost impossible if they are mixed at the press using a beam scale. Provide your ink supplier with an approved customer color sample and samples of the paper the job will run on. Using a production printing press as a proof can be very expensive.

Blankets

Choose a blanket based on print quality and dot gain characteristics. Document the effectiveness of new blankets over a specified period of time to keep track of performance for future purchasing. This data will also help you determine the origins of poor print quality, blanket problems, and potential machine deficiencies.

Specifications

Blanket specifications should include the following:
- Manufacturer/brand name
- Type (conventional or compressible)
- Thickness (3-, 4-, or 5-ply; variance no more than ±0.001 in. across the entire surface of each blanket)
- Size and squareness (many printers specify a different size than the manufacturer recommends to better fit the needs of the press)
- Blanket bar specifications (hole-punched, non-hole-punched, premounted aluminum bars)
- Process application (conventional sheetfed, heatset or non-heatset web, newspaper, UV, EB)
- Delivery date and destination
- Shipping instructions (same day, overnight, freight)

Improper blanket storage can result in the blanket becoming imprinted with the texture of the back side.

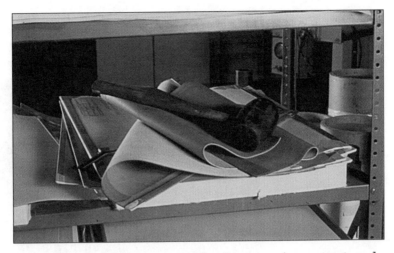

Thickness tolerance specifications are important and should be developed when consulting with the supplier. The thickness of a blanket should vary by no more than ±0.001 in. across the entire area of each blanket. (Make sure this specification appears on all purchase orders.)

Inspection

Blanket thickness should be measured with a deadweight bench micrometer in at least nine points across the blanket. Reject any blanket out of tolerance, and send it back to the supplier for replacement.

Handling and Storage

Blankets should always be stored in cool, dark areas that are unaffected by heat. Always be sure that the storage area is free from ozone that emits from sparking electric motors. These emissions oxidize the blanket, which will cause streaking on press. Because of the environmental factors affecting blanket quality, it is best that they remain in the blanket tube until they are ready to be used. If blanket bars are mounted for blanket staging, they still should be protected from environmental effects.

Rollers

Choosing the right rollers and properly maintaining them is essential, especially if you are using alcohol-free fountain solutions or the waterless process. Much like blankets, rollers deteriorate over time and must be replaced.

Specifications

Consult with your supplier when establishing the following specifications:
- Application (sheetfed, heatset or non-heatset web)
- Type (conventional inks, UV inks, heatset inks)

Check that all blankets are of the appropriate thickness, using a deadweight bench micrometer *(right)*.

- Conventional lithographic or waterless
- Roller type/location (ink form, distribution, dampening form, dampening metering)
- Press manufacturer specifications
- Roller hardness
- Roller dimensions
- Run-out tolerance (degree of chamfer at the roller edges)
- Oscillating ink form rollers
- Shipping requirements (UV light protection and damage-resistant wrapper)
- Delivery date and location

Inspection

Rollers should be inspected on delivery for compliance to the purchase order, specified requirements, and visual damage. The hardness and dimensions of the rollers should be measured with a type A durometer and special measuring rule.

Handling and Storage

Rollers need to be stored vertically with their protective wrappers intact, and they need to be protected from UV light sources. If rollers are not properly maintained, glazing of the roller surface will quickly develop. A glazed roller surface will not carry as much ink as a non-glazed roller (much the same as the way a new paintbrush will hold and transfer the paint much better than a brush with previously dried paint). A glazed roller surface will look shiny and the surface will be smooth. Durometer measurements will also be much higher than when the rollers were new. Establishing a realistic roller maintenance program will help ensure satisfactory roller life.

Measure roller hardness with a durometer *(left)*.

Proper storage for rollers is vertically with wrappers intact *(right)*.

 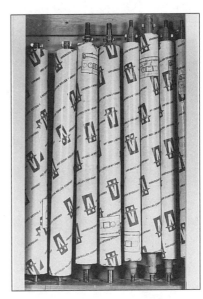

Summary

Controlling the quality of pressroom materials starts with establishing a sound quality assurance system with choosing suppliers based on:
- The quality of their products
- The extent of their quality system and customer service
- Their willingness to participate in your quality efforts
- Their ability to certify that their products will consistently meet your specifications

Taking steps to control the quality of your pressroom materials can minimize or eliminate problems down the line on press, saving you time and money and helping you meet customer expectations of high quality and service.

6   Technical Systems Control

Technical systems control is an essential element in the printer's seemingly endless search for ways to provide quality products and services. "Technical systems" include all the processes, materials, and equipment required to produce the printed product from start to finish. The "control" aspect has three parts:

1. **Compiling the checklist**
    Identify every component that requires control, from pre-press to pressroom and finishing.

2. **Measuring and testing the components**
    Determine if the main technical components are compatible and if they are operating as specified, based on the checklist that was developed.

3. **Conducting the audit**
    Conduct periodic and structured evaluations of the technical system to determine if the measurement and testing based on the checklist are being carried out as specified and are operating correctly. The first part is an internal audit; the second is a third-party technical systems specification audit.

Compiling the Checklist

Your checklists may differ from the examples on the following pages, but the principles remain the same. This section concentrates on developing the checklist. Later sections will cover other aspects of technical systems control.

Pre-production note. Because job specifications and related information must be accurate before a job is put into production planning and scheduling, it is wise to review and verify all the pertinent facts. Simple clerical errors can easily compound, especially if every person along the production trail re-interprets the error.

Prepress Checklist

☐ **Work area environment.** Are the following conditions being controlled, measured, and checked continually: air temperature, relative humidity, static control, positive air pressure, daily and weekly housekeeping, department doors being kept closed, and degree of dirt/dust contamination from outside the prepress area?

☐ **Department lighting.** Is safe lighting in camera and plate areas being checked to prevent pre-exposure of films and plates? These areas need red or yellow safelights and/or indirect UV-shielded white light conditions. Potential white lighting contamination of film from the main prepress areas should also be checked.

☐ **Incoming consumable materials.** Are materials and chemistry being checked on arrival for packaging integrity or possible shipping damage and for compliance to invoice specifications? New film and plate shipments should be checked with certified test targets to determine if they meet specifications for exposure and resolution.

☐ **Film output equipment.** Are film imagesetters and film processors, cameras (horizontal and/or vertical), contact frames, and step-and-repeat cameras being checked for proper and consistent exposure, accurate dot reproduction, and register?

☐ **Film punches.** Are film punches, both stand-alone types and those internal to image-setters, being checked for consistency and accuracy?

☐ **Film flats.** Are film and film flats—including those produced in house or by the customer—being inspected for internal fit and image assembly register accuracy prior to platemaking? Such inspections can help detect if damage has occurred during storage and if the storage facility and its environment are adequate.

☐ **Proofing systems.** Are certified test targets being used to check consistent dot reproduction and compatible dot gain characteristics on each proof?

☐ **Stripping and layout tables.** Are the edges of the stripping and layout tables being checked weekly to make sure they are square. Is the T-square really "square"? Is the surface of the stripping table clean and free of bits of tape and other debris?

☐ **Plate punch.** Has the plate punch been inspected for accuracy, including squareness of the die punch, self centering, squareness of the gripper edge, and compatibility of the punch dies to the pins.

☐ **Platemaking equipment.** Are certified test targets being used at specified intervals to verify that exposure, dot reproduction, and register are accurate and consistent? Are the correct type of register pins in the platemaker? Plate processing consistency also depends on regularly checking for machine mechanical problems and checking machine chemistry maintenance. These checks help ensure that accurate and consistent plates are going to press.

☐ **Plate inspection and handling.** Has a plate inspection station been installed? Does it have proper lighting, a large magnifier, and an accurate straight edge? Can the plate inspector conduct inspections with minimal physical discomfort? Does the plate inspector have a procedural checklist to facilitate conducting the inspection?

Press Checklist

☐ **Work area environment.** Are the following conditions being regularly checked and monitored: temperature and relative humidity, locations of heating and air conditioning vents and their influence on press units, lighting conditions, and housekeeping in the general area and on press?

☐ **Color viewing conditions.** Are color viewing conditions, including the color temperature of the bulbs and their illumination and surrounding area, based on established ANSI standards? Are bulb temperature and illumination being measured regularly to make sure they adhere to ANSI standards?

☐ **Incoming pressroom materials.** Are pressroom materials and supplies being purchased based on quality and compatibility, not just price? Are materials being inspected for possible shipping damage and for conformance to invoice specifications? Is required data from the supplier (e.g., paper weight, thickness) gathered internally by the printer and evaluated to determine quality and compatibility? Are storage facilities required for maintaining the integrity of the materials being regularly inspected?

☐ **Press preventive maintenance.** Has a realistic preventive maintenance program been established and is it being carried out regularly? Have daily, weekly, and monthly checklists been developed based on manufacturer specifications? Are completed checklist documents being kept on file for future reference? Are machine breakdowns and maintenance data being regularly reviewed so you know your preventive maintenance program is actually working?

☐ **Printing pressures.** Are the correct plate/blanket squeeze and impression cylinder pressures used according to industry/manufacturer specifications? Are the following being checked with industry-accepted instruments: cylinder bearer contact or gap; bearer height for plates and blankets; thicknesses of plates, blankets, and packing; and impression cylinder pressure settings?

☐ **Rollers.** Are the proper shore hardness and press settings being checked and maintained on a regular basis? Roller settings should be based on the press manufacturer specifications. Shore hardness should be measured with the appropriate durometer.

☐ **Fountain solution.** Have the correct pH and conductivity been established? Are they being monitored during production? Do charted and documented measurement data include pH, conductivity, and water temperature? This can help determine what changes to make if dampening problems occur.

☐ **Technical system analysis.** Is a comprehensive press test form (e.g., GATF Sheetfed Test Form) being used to conduct a twice yearly press analysis? Is the test being used to (1) diagnose potential mechanical problems, (2) determine run control capabilities (densities and dot gains) with specified materials, and (3) calibrate the press to establish density and dot gain specifications for prepress color separations.

☐ **Density and dot gain control.** Are calibrated density/dot gain measurement instruments and quality control targets (color bars) being used to monitor print sheet quality throughout the pressrun? Have quality control procedures and a system of customer OK's been established? Are they being used consistently?

Verifying, Measuring, and Testing

The previous section explained the first step in technical systems control in print production—compiling the checklist. This section describes what to do next—performing the testing, measurement, and verification procedures based on the checklist.

To be effective, testing and measurement need to include checklists; accompanying records should be kept consistently and accurately; and testing and measurement should be done on a specified schedule—weekly, monthly, or quarterly—for equipment and on receipt for materials. Saving shipping invoices, packaging labels, and individual packaging identifications is essential to controlling the materials. The testing and measurement procedures, their frequency, and work instructions should all be documented.

Having a structured quality system is the best way to ensure that procedures are properly developed, implemented, and maintained. The most recognized quality system worldwide is the ISO 9000 series of quality systems standards. Of the twenty system clauses in the standard, six focus on measuring, testing, and data recording for controlling the main components of a technical system.

Conducting the prepress and press checks identified here will help maximize the productivity of all your equipment—and your entire technical system—and ensure that you produce the highest possible quality for your customers.

Prepress

❏ Work Area Environment
1. Maintain temperature and relative humidity at optimum conditions (70–75°F and 45–55% RH) to minimize static. Ideally, a certified instrument continuously monitors and records conditions for daily evaluation.
2. Maintain positive air pressure in the prepress area to help minimize dust, paper lint, and spray-powder contamination on film, plates, and equipment. Check air flow daily by opening the doors slightly to see if air is entering (negative flow) or leaving (positive flow).
3. Prepress area floors are tile covered or sealed concrete. Wet-mop the floor daily to help remove dust.

❏ Department Lighting
1. Install safelighting to prevent premature exposure of film and plates. Check safelight guide filters (blue for yellow lighting and green for red safelighting) when the bulbs are installed and once a month thereafter.

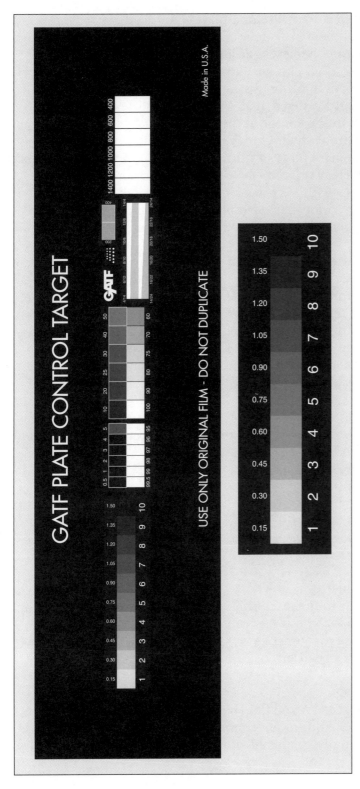

The GATF Plate Control Target (*top*, order no. 7120) is imaged along with the production work to monitor the platemaking or imaging process. The target can be used to test the uniformity of illumination in a vacuum frame by assembling several targets on a flat so that one images in each corner of the frame and one in the center. Illumination differences are seen as differences in the exposed images of the targets. The exposure level given to the plate is indicated by the critical step that is found on the continuous-tone scale (*bottom*). The exposure recommendations of the plate manufacturer should be followed. For many plates, a critical step 4 or 5 will be recommended.

Contact your film and plate manufacturer for information about the correct safelight bulbs and materials.

2. Conduct a fog test at least once a month. Place an unexposed plate or film where it will be most exposed to room lighting during production. Cover the plate except for three inches of an outside edge, placing a coin in the center of the uncovered area. Move the cover about three inches every two minutes to expose the entire plate of film to room safelighting conditions. If a clear area the size of the coin appears when the plate or film is processed, fogging conditions exist. Note: Aqueous based plates appear more sensitive to fogging when improper yellow or white lighting is used.

❐ Incoming Consumable Material and Film Output Equipment
1. Inspect invoices and packaging; verify that they meet documented specifications.
2. Test film and plates for proper exposure times and processing using a certified test target (the GATF exposure target shown on page 87) before using them in production jobs. Test film exposure equipment to determine if it is accurately and consistently reproducing film.

❐ Proofing Systems
1. Make sure each color proof, both internal and supplied, has a test target (for example, the GCA/GATF Proof Comparator III shown on the opposite page) exposed in the proof.
2. Measure the proofing target with a Status-T reflection densitometer for density, dot gain, and hue error and grayness. Document data so you can compare for accuracy and consistency.

❐ Film Punches
1. Accurately match film punch die hole sizes to the pins used for image assembly.
2. Check die punches weekly; make sure they are in proper position, not damaged.

❐ Film Flats
1. Inspect film and flats before platemaking or proofing.
2. Check all film for accurate internal fit, accurate image assembly, damaged stored film, and assembly integrity.

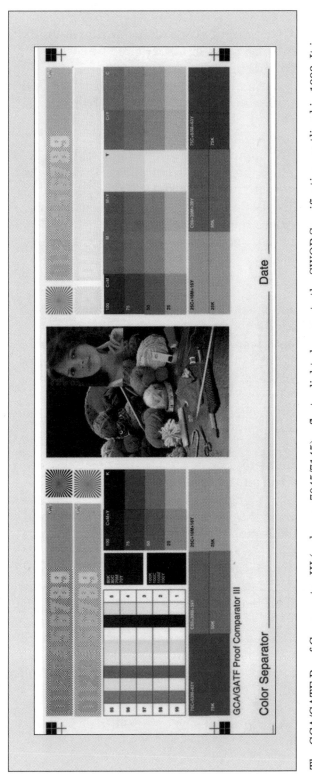

The GCA/GATF Proof Comparator III (order no. 7045/7145) reflects slight changes to the SWOP Specifications outlined in 1993. It is a film target that evaluates the accuracy and consistency of off-press proofs. The Comparator consists of a set of separation films, a proof of which becomes a reference comparator against which the Proof Comparator images on production proofs are evaluated. The Proof Comparator III measures 1.8×6.3 in., with a screen ruling of 133 lpi. Key elements include a pictorial centerpiece contain a wide variety of colors and textures; solids, tints, and overprint patches, gray balance bars, star targets, a dot gain scale, slur targets, exposure control patches, resolution blocks, and three-color gray patches.

3. Maintain film storage area within accepted temperature and relative humidity conditions for film stability.
4. Perform an in-house proof of all supplied film, then compare it to proofs provided with films.
5. Include a scale with all film (camera, imagesetter, and dupe) so accuracy can be measured with a transmission densitometer before platemaking.

❐ Image Assembly (Stripping) and Layout Light Tables
1. Check light tables weekly for squareness. To check, line up and rule out both sides of a sheet on the light table using a certified T-square. If the ruled lines don't line up, the table edges require correction.
2. Check illumination surfaces weekly for chips, cracks, and other damage.

❐ Plate Punch
1. Check plate punch squareness twice monthly. To check, punch two plates separately. Place them face to face, then put stripping pins in the center and left die holes. Use a 25× magnifier to see if the opposite die holes and gripper edge of the plate line up. If they don't, the plates and/or the images will not be square.
2. Make sure die-punched holes in the plate are fitted to the pins in platemaking and in the pressroom.

❐ Platemaking Equipment
1. Check pins in the plate exposure equipment. They should be either spring-loaded in a pin board or small-based solid pins with plastic tabs. Make sure step-and-repeat equipment pins are square and snug so there is no movement.
2. Test vacuum frames and step-and-repeat equipment for register once a month. (A Register Test Grid is available from GATF.) Four burns at each step should be dot to dot ±0.0005 in.
3. Test vacuum frames and step-and-repeat equipment for exposure accuracy at least twice a month. Test the center and four corners on a vacuum frame for light falloff. Expose a minimum of six steps on step-and-repeat equipment. Consult with your plate manufacturer about the proper continuous-tone exposure and micro-line resolution.

Register test grids, available from GATF (order no. 7060), aid in assessing register of vacuum frames and step-and-repeat machines, evaluate image fit across the press, test consistency of register, and detect paper fan-out.

❒ Plate Inspection and Handling

1. Identify plates with color, job title and number, and inspect them against a checklist that enumerates all instances of nonconformity. The checklist (shown in chapter 3)—dated, showing items checked off, and including inspector's initials—should go with the job jacket.

2. Install lighting that conforms with ANSI standards in the inspection station. Include a large magnifier; make sure the station is designed to minimize discomfort.

3. Use a plate remake form for all remakes. A form that lists all possible problems to check off will enable Pareto charting to quickly track down particular plate problems.

Press

❒ Press Work Area Environment

1. Maintain temperature and relative humidity at optimum conditions (70–75°F and 45–55% RH) to minimize static, help maintain paper stability, and minimize potential fountain solution, plate, and ink problems. Again, a certified instrument should continuously monitor and record conditions for daily evaluation.

2. Check press area lighting with a light meter once a month. Make sure the lighting is 75 footcandles between the press units at the level of the plate cylinder.

3. Visually check basic safety housekeeping and document daily. Check for oil leaks, water and fountain solution spills, proper tool storage, air pressure leaks, waste-paper, and general cleanliness.

❑ Incoming Press Materials
1. Make sure inks meet specifications and that the supplier has provided the following record keeping data for each ink batch shipped: order verification, batch number and ink tack at test RPMs, water pickup of the ink batch received, spectrophotometer measurements, Little Joe proof of each ink on the substrate to be printed at 0.2–0.4 mil ink film thickness for process inks and up to 0.5 mil for PANTONE® and special matches.
2. Review, record, and file all paper mill documents. Verify that the paper received matches order specifications. Also check mill out-turns, moisture specifications, caliper tolerances, and label information.
3. Verify rollers against specifications, including (a) invoice, packaging conditions, and label data, (b) actual length and circumference versus manufacturer specifications, and (c) shore hardness (durometer measurement). Normally, new roller shore hardness measurements for ink and water forms are 22–25; for water metering rollers, 18–22, and for ink distribution rollers, 28–30. Make sure you consult the press manufacturer's specifications.
4. Store rollers vertically.
5. Maintain temperature in roller storage area at 65–80°F.
6. Verify that blankets received match specifications, including invoice and label data. Verify blanket caliper against specifications stamped on blanket. Use a Cady gauge to take measurements at nine locations; measurements should be ±0.001 in. of specs.
7. Check that premounted blanket bars are properly secured to blankets.

❑ Color Viewing
1. Maintain all press sheet and color viewing locations within ANSI color viewing conditions. Check lighting twice a month to make sure (a) the lighting color temperature is 5000K (GATF/RHEM Light Indicator); (b) the lighting illuminance is 165 to 210 footcandles (light meter); (c) the color-rendering index is ≥90; and (d) the surround is Munsell N/8 matte gray.

Standard color viewing booth.

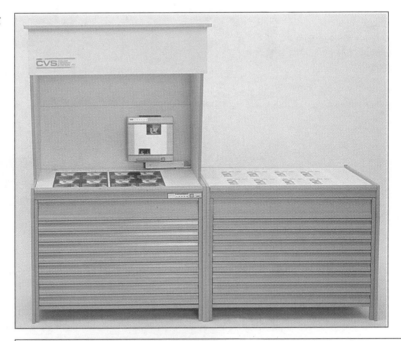

GATF/RHEM Light Indicator (order no. 7065-R), used to signal to the viewer whether or not lighting is at standard 5000K.

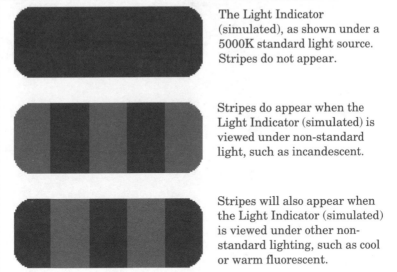

The Light Indicator (simulated), as shown under a 5000K standard light source. Stripes do not appear.

Stripes do appear when the Light Indicator (simulated) is viewed under non-standard light, such as incandescent.

Stripes will also appear when the Light Indicator (simulated) is viewed under other non-standard lighting, such as cool or warm fluorescent.

❏ Printing Pressures
   1. Check bearer-to-bearer ring contact at least once a month. For sheetfed presses, you can use the ink thumbprint test at six locations around each bearer ring. For web presses, use the visual light test or the aluminum foil impression test. For presses that run off bearer, use feeler gauges according to the press manufacturer's specifications.

A blanket packing gauge is used to check plate-to-blanket squeeze.

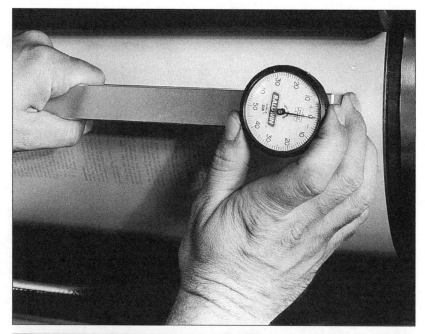

Measuring blanket caliper with a Cady gauge.

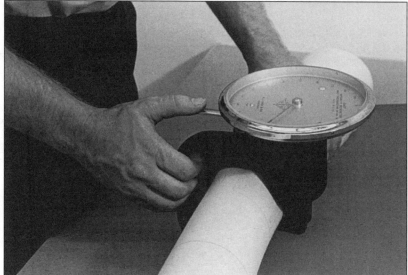

2. Check the plate-to-blanket squeeze, using an accurate packing gauge, each shift and when a new blanket is mounted on the cylinder. The total squeeze between plate and blanket should range from 0.004 to 0.006 in. over bearer height for most compressible blankets and from 0.002 to 0.004 in. for conventional blankets.
3. Record data in press data log.

❐ Rollers
1. Check roller settings to the plate twice a week and record them in the TPM log. (One way to document the settings is to print the roller stripes on a few clean sheets.) Most roller companies supply accurate measuring gauges. Roller settings should be maintained at the press manufacturer's specifications.
2. Check the rollers' shore hardness once every six months (part of the TPM program) with a type A durometer. Replace a roller when the shore hardness is more than ten points above the shore hardness when it was new. Visually check rollers for surface glazing and deterioration when you check the shore hardness.
3. Record the results in a roller maintenance log.

❐ Fountain Solution
1. Using an accurate meter, measure fountain solution chemistry every shift the press runs. Record data on a control chart and monitor to determine if changes occur with the water, at press, or in the automix system.
2. Measure pH and conductivity of both water and fountain solution. Base the fountain solution targets on the best concentration established in the pressroom.
3. Take corrective action if the pH changes more than one number during a shift. A change from 4.2 to 5.2 equals 10 times less acidity.
4. Closely monitor print quality when a conductivity change of more the 500 micromhos occurs within a shift. If the water feed on press must be increased more then 20%, corrective action may be needed.

❐ Press Optimization. Optimizing your press to its highest capabilities is a three-stage operation that includes diagnostics, a capability study, and calibration.
1. Diagnose and correct any mechanical and technical press problems.
2. Determine upper and lower density variations through a pressrun.
3. Establish density and dot gain targets after determining maximum densities through print contrast calculations.
4. Use a test form (GATF has several, including the Newspaper Test Form shown on the following page) to properly conduct your press optimization study.

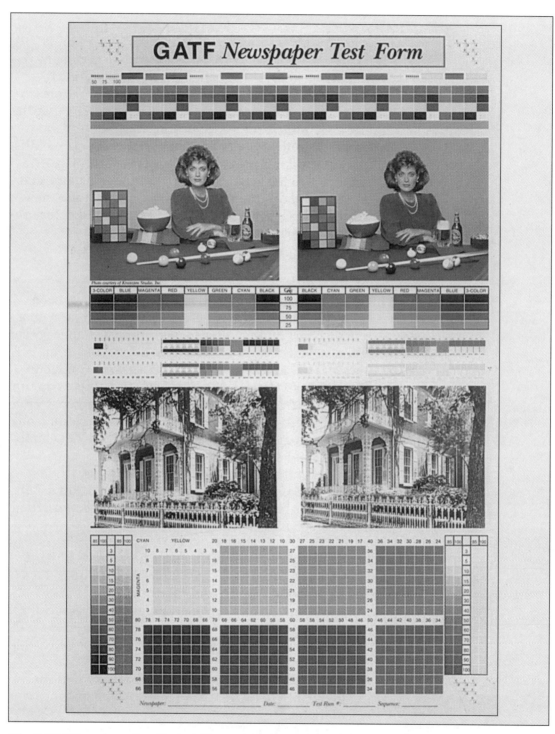

The GATF Newspaper Test Form (order no. 7091/7191) is used to determine print characteristics of newspaper presses, including gray balance, dot gain, print contrast, tone reproduction, and trap.

5. Conduct a press optimization study twice a year to ensure that your presses are providing the best possible quality for your customers.

❒ Density and Dot Gain Control. Density and dot gain control require tools beyond common visual analysis of the pressrun. These tools include a calibrated Status-T densitometer and original, certified color control bars. Implement quality run procedures to ensure consistent density and dot gain throughout the pressruns.
1. Run and monitor jobs to established density and dot gain targets. Dot gain for sheetfed litho can be around 18% ($\pm$4%), and for heatset web 24% ($\pm$4%).
2. Measure densities with a Status-T densitometer every 500 to 1,000 sheets through the pressruns.
3. Measure dot gains periodically throughout pressruns.
4. Visually monitor the color control bar (gray balance patches, star targets, dot gain scale targets) to look for causes of possible shifts in hue of the cyan, magenta, and yellow inks.
5. Follow recognized specifications for density and dot gain control—SWOP (Specifications for Web Offset Publications) for commercial sheetfed litho and heatset web presses; PROP (Prepress Recommendations for Offset Packaging, type 1 folding carton) for sheetfed litho folding carton printers.

Total Production Maintenance

A realistic, structured total production maintenance (TPM) program is recommended for all production equipment. Consulting with suppliers and manufacturers is essential when developing a TPM program.

Maintenance logs and checklists, either hard copy or computerized, are the main mechanisms for verifying the effectiveness of a TPM program.
- Document and maintain records for your daily, weekly, and monthly checklists.
- Generate monthly SPC charts (Pareto, flowcharts, and control charts) from the TPM checklists.
- Determine the effectiveness of the TPM program by putting the SPC chart under a management review.

Auditing the Results and Procedures

As explained in the first section of this chapter, "technical systems" include all the processes, materials, and equipment required to produce the printed product from start to finish. This section focuses on the audit, a comprehensive, periodic structured evaluation of the technical system.

What Is an Audit?

An audit of the technical system is a systematic examination to see if the technical system is producing consistent quality results that comply with manufacturers' specifications for equipment and materials. Audits verify that technical system components are operating as designed and producing products of consistent quality.

There are two primary types of audits: (1) an internal audit, conducted by company personnel, and (2) a third-party technical systems specification audit, performed by professional auditors/consultants. Periodically conducting both types of audits ensures that the technical system remains in control and is continuously improved. An audit's primary purposes are to:

- Meet customer quality requirements
- Determine if the technical system elements conform with manufacturer and industry-accepted specifications and accuracy
- Determine the effectiveness of the technical systems quality operating procedures
- Verify that the technical system checks are accurate and are being conducted regularly
- Give the department being audited the opportunity to improve its technical systems components

Basic audit forms should be developed to provide consistent documentation and to make it easy for auditors to document the test and audit results.

Auditor Responsibilities

Internal auditors can come from middle management, staff, or production areas and they should have basic knowledge about the technical system components. They should also have training in how the tests and checks are conducted and how to review the documents that record the testing, measuring, and verification results.

Third-party audits verify the technical system's performance and internal audit effectiveness. They should be conducted by experienced auditors who are not affiliated with the printer. Third-party auditors who are specially trained to conduct audits can not only offer unbiased and independent analyses, they can also provide insight into benchmarks and trends in the graphic communications industry.

Internal auditors need to conduct themselves in the same way as external auditors—with professionalism, independence, and objectivity. Auditors are responsible for the following:

- Complying with the defined audit requirements of the technical system
- Planning and conducting the audits
- Documenting observations made during the audit
- Clearly conveying the requirements of the technical system checks during the audits
- Reporting the audit results to an audit review committee
- Following up and verifying the effectiveness of corrective actions

Audit Frequency   Audit frequency should be based on defined quality requirements and on how often technical checks are conducted in the production departments. For example, prepress personnel may test the performance of cameras and imagesetters once a day or once a week, but they may check vacuum frames and step-and-repeat equipment only once a month. Press operators might measure density/dot gain at least once during each pressrun, the pH and conductivity of dampening solution daily, and form roller settings twice a week. Lighting conditions (for both work area and color viewing) may be checked monthly. Many printers conduct press diagnostics twice yearly.

Summary   As a general guideline, internal audits should be conducted at least twice a year. This allows enough time for plant personnel to perform technical system testing and to document the results so that the auditing process will be effective. An external third-party audit should be conducted approximately once a year.

7 Total Prepress Maintenance
by Ron Bertolina and Chuck Koehler

Manufacturers of prepress equipment will recommend procedures and schedules for maintaining their products, and
those should be strictly followed to maintain optimum prepress productivity, reduce or eliminate prepress downtime,
and avoid safety hazards. Improper or negligent maintenance
and repair can often result in any warranties that may cover
equipment being voided. Some prepress departments establish service agreements for regular maintenance, but they
still need to perform daily maintenance in many cases.

Many of the conventional equipment maintenance checkpoints also apply to digital equipment (as in the case of plate
processors). In a high-volume shop, where processors are
used around the clock, maintenance is as important a routine as is any production process. In small-volume shops,
where processors are not run around the clock, it is just as
important and should not be ignored—perhaps it is even
more important because the equipment sits idle and parts
end up sticking together. There should be ample workspace
around any machine so that heat can dissipate, and so maintenance on the machine can be performed unhampered. The
environment in which all equipment is placed should be followed according to manufacturer's specifications.

In some cases, depending on the features of your equipment, the checkpoints listed in this chapter may not apply.
But what does apply in all cases of maintenance is documentation of (1) what needs to be done and when to do it; and (2)
what was done, when it was done, by whom, and any comments about the procedure that management or fellow
employees should know about.

Most equipment requires calibration procedures as
instructed in user guides. Some calibration procedures
require special software. Calibration, along with mainte-

nance, helps to keep output linear (when you ask for a 20% dot, you get a 20% dot).

Continually check warning lights on all devices that prompt you for maintenance. When problems arise, it is best to consult the operator's guide for possible solutions to many problems. Troubleshooting remedies should be documented as thoroughly as maintenance documentation so that all involved may learn, and a maintenance history can be tracked.

And, by all means, keep safety in mind when doing any maintenance or repairs. Be aware of Material Safety Data Sheets (MSDS) on handling chemistry, wear appropriate glasses and clothing, know where the nearest eyewash station is, and never attempt to repair anything electrical unless you are properly trained and qualified to do so. Never attempt to defeat or disengage any safety precautions. Always shut off the main power supply to any equipment when doing maintenance or repairs.

Prepress Production Rooms

☐ Have floors mopped or vacuumed every night; keep dirt and dust to a minimum.
☐ Keep positive air pressure. Dirt and dust is forced out, not pulled into the room (as with exhaust systems). Filtered air is suggested.
☐ Maintain 50% relative humidity without condensation; maintain temperature between 68–75°F.
☐ Use antistatic devices, static eliminating equipment, Dust Bunnies, and Impress.
☐ Check daily for leaks on all processors; repair immediately.
☐ Report any switch malfunction to a supervisor and fix immediately.
☐ Dispose of waste chemicals promptly and responsibly.
☐ Never use any harsh cleaning compounds, which may void the warranty of equipment.
☐ Never use an abrasive that could damage equipment.
☐ Always refer to equipment manuals before doing any formal maintenance or repair of equipment; never try to repair equipment if doing so would void a warranty.
☐ Use safelighting to prevent fogging of light-sensitive materials while handling and processing. Some safelight filters are fade-free for long life. Where safelighting is required, they should be regularly checked for adequacy; use a safelight test guide and follow instructions.
☐ Use voltage regulators to help during power fluctuations.

❏ Keep spare fuses on hand for all applicable machines; always use the proper replacement fuse.
❏ Know where all emergency power shutoff switches are located.

Color Scanners

❏ Check laser voltages according to manufacturer suggestions
❏ Check and clean film exposure drums regularly
❏ Clean and oil lead screws according to manufacturer
❏ Clean or empty film punches as necessary
❏ Change scanning bulbs as needed or recommended
❏ Do not touch the bulb with your fingers
❏ Inspect and clean light arm attachments for CCD scanners
❏ Clean and inspect flatbed glass panels regularly
❏ Clean acrylic drums regularly and inspect for scratches
❏ Use appropriate cleaning materials
❏ Polish acrylic drums as necessary with approved polish
❏ Store drums according to manufacturer recommendation
❏ Install software according to manufacturer's instructions
❏ Install program upgrades as they become available
❏ Reload software at any sign of corruption
❏ Store software in a dry place away from magnetic fields
❏ Make backups of system or operating software and store accordingly
❏ Make and store setup files as necessary
❏ Maintain materials and equipment according to manufacturer's specifications:
  ○ Copy scaling and calculation
    • Film type ruler
    • Electronic calculator
  ○ Copy preparation and mounting
    • Appropriate film cleaner
    • Adequate wiping products

Drum scanner.

Use the appropriate cleaner to remove dirt, lint, and foreign particles from the surface of the copy and from the scanner drum or bed.

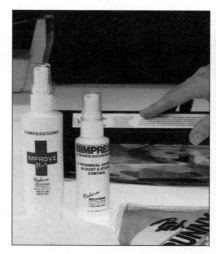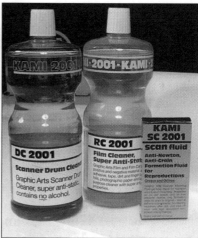

- Clear tape (various widths sometimes are beneficial)
- Metal straight-edge or ruler
- Anti-Newton rings product (spray or powder)
- Mounting gel or oil
- Mounting acetate (if required)
  ○ Maintenance/cleaning
    - Proper drum cleaning fluid
    - Proper drum cleaning cloths or tissues
    - Proper drum polish
    - Proper lubricating fluids
    - Appropriate tool kit

Light Tables

❏ Square light tables regularly
❏ Clean glass
❏ Replace bulbs when needed; keep several on hand

Squared light table showing assorted stripping tools.

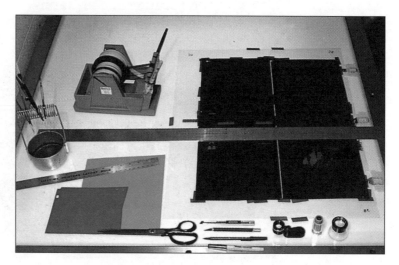

Vacuum Frames

❏ Use a register test grid regularly to check register accuracy of vacuum frames.

❏ Flush out vacuum pumps; all vacuum ports and hoses should be clean and unobstructed; hoses and connections should be tight; inspect hoses for cracks (leakage). With dirty vacuum pumps, vanes in the pump can stick and cause uneven drawdown from one exposure to another. Dirty filters restrict air cooling, allowing the pump to heat up and carbonize the oil.

❏ Clean frame blanket of debris using a vacuum cleaner.

❏ Use a non-spotting, residue-free glass cleaner to clean dirt particles from glass.

❏ Use static eliminators, Dust Bunnies, and Impress products; ground the vacuum frame.

❏ Maintain the manufacturer's recommended vacuum pressure, usually referred to in inches of Hg (Mercury); generally, vacuum time should be one minute after Newton rings appear evenly over the entire frame.

❏ Use an integrator with a photocell to regulate exposure variations caused by aging lamps. Keep photocell dust free.

❏ Use a UV degradation device that can monitor UV output quality.

❏ Replace bulbs when manufacturer's recommended bulb life ends, not when bulbs burn out. Bulbs generally last 1,000 hours. The quality of UV output dramatically decreases as the bulb ages. Timely replacing them will help to ensure proper platemaking. Develop plate count

Overhead view of a vacuum frame in operation.

logs to help in this regard— how many plates are you running through the processor at any given time? Get counters to help determine when bulbs need to be replaced.

❐ Test a bulb's effectiveness by exposing a plate to its recommended step and run a plate through the processor, using fresh chemistry. Take a developer-soaked pad and rub over the scale. If the scale drastically drops, the bulb is suspected to be weak.

❐ Check for even illumination by exposing a step wedge to a plate in all corners and the center of plate. Evaluate scales to determine exposure consistencies over entire plate area.

Film D_{max} = maximum density, most recommend 3.5–4.5

Dark or black areas of film. Avoid "burn through" (beyond 4.5) to keep sensitive serifs or other thin elements from pinching up. As a habit, measure the same spot on every film for consistency.

Film D_{min} = minimum density ranging 0.03–0.07

Clear areas of film. If you get readings higher than 0.07, you may have contaminated fix (too much developer in fixer); need to avoid "dichroic fog," which is a milky-looking or flat-looking overall film appearance. This dichroic fog will clear once you run the film back through the processor filled with uncontaminated fix. Another sign of contaminated fix is the smell of rotten eggs.

Fill out a density monitoring chart. If D_{max} and D_{min} start veering off course, it can be caught before it gets out of control. This is historical documentation for a service technician to better know what is happening so that it isn't such a "blind" troubleshoot. When calling for technical assistance for film processor problems, it is recommended that you have the following information handy:
- The film and chemistry used; evidence of the film (do not trash)
- Development time (actual, not what the dial reads)
- Development temperature (actual, not what the dial reads)
- Fixer temperature (actual, not what the dial reads); use a bi-metal thermometer
- Dryer temperature (actual, not what the dial reads)
- Control strip values if used
- Make, model, and size of processor being used

Special control test strips monitor processor behavior. Because control strips can be expensive, checking the film D_{min} and D_{max} often helps in catching processing problems and is an alternative to control strips.

❏ Clean rubber seals ("beads"); they should be soft and resilient; a blanket rejuvenator or silicone rubber cleaner can be used. Air leaks from a cracked rubber seal will slow drawdown and prevent intimate contact between materials.

❏ Clean exhaust and fan; keep them unobstructed. On some frames (especially flip-top frames), heat buildup can be excessive, causing film expansion.

❏ Keep reflectors on exposure lights clean.

❏ Maintain contacting boards; avoid using pin bars or stripping pins in vacuum frame boards that will cause distortion in contacting and divots in the boards; use small-base pins in vacuum boards that are not too high or too short to avoid misregister problems; replace or repair boards if they have cracks near the edges of punch holes.

❏ Replace film step wedges often because the actual silver emulsions deteriorate and give false results.

Punch Systems

❏ Check regularly for integrity of punched holes. Worn punches should be replaced immediately

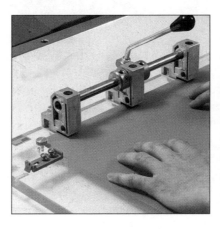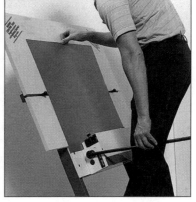

Densitometers

❏ Calibrate daily. Use a transmission densitometer to assist in plotting and monitoring film D_{min} and D_{max} behavior several times during each shift to keep a problem from getting out of control. Helps in troubleshooting

❏ Use a sequence to read film density: zero straight to bulb; measure D_{min} on clear area; zero again to bulb; measure D_{max}

❏ Always read the same area of the film (for consistency)

Step-and-Repeat Equipment

❏ Use a register test grid regularly to check register behavior
❏ Replace aged bulbs before they burn out
❏ Use a step wedge on every plate to insure proper exposure, check resolution targets

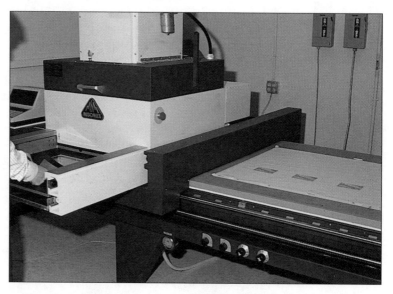

Film Processors

❏ Clean thoroughly on a regular basis; remove any crystallization on rollers and guides; never use any harsh abrasive that may damage rollers (plate processors, too)
❏ Keep filled with uncontaminated, non-exhausted chemistry; use stirring paddles for mixing
❏ When mixing chemistry, always follow the recommended ratios: example, mix four gallons of developer to one gallon of water (4:1). The first number in the ratio should always refer to the chemistry, the second number is the water amount. When mixing new chemistry, drain and clean out the old tank and reservoir (get rid of any old chemistry laying at the bottom because this will defeat the freshness of the new chemistry). When mixing chemistry, stir well, especially with fix to avoid a layering effect that could occur when fix is not mixed well. Never add hardener to fix without water first being in the solution (follow instructions). Mix fix and add it into the processor first before mixing and adding developer (for danger of contaminating developer) use a splash guard over the developer tank when adding fix into the tank.
❏ If you are mixing chemistry and forget what you've already mixed, use pH strips to determine the acidity instead of dumping the chemistry and restarting.

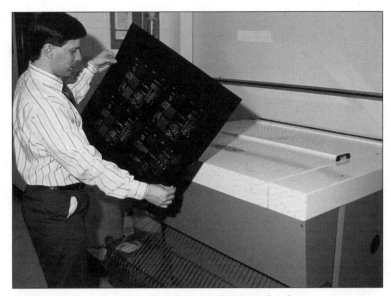

❒ Set and maintain developer, fix, wash, dryer temperatures and speed controls according to manufacturer's recommendations (use proper thermometers to verify temperatures; stop watches to verify times)

❒ Do not let film dryer temperature become so high that it interferes with film stability; 115°F is too high

❒ Check replenishment rates for all chemistry on a daily basis: replenishment rate is the quantity of chemistry delivered to the machine to make up for the amount of chemistry being used for each media being processed

❒ Check oxidation replenishment when processor has been left in stand-by mode for a long period

❒ Use lids on chemistry—cuts down on oxidation

❒ Replace all filters regularly and clean out filter housings

❒ Maintain chemistry capacity in tanks

❒ Use the correct chemistry for the media being used

❒ Regularly drain wash tanks to prevent scum and algae from forming; to avoid algae, drain often and dry; algae grows because of water. Photobromes are toxic to algae and deter algae growth.

❒ Check all rollers and gearbox for signs of wear, excess play, pressure consistency

❒ Check chain alignment, tension

❒ Check fixings and fasteners wherever accessible

❒ Lubricate steel worms on the mainshaft to minimize wear on worm gears; silicone grease is commonly used

❏ Check recirculation—this is usually when uneven development shows up; clear clogged roller racks; check dirty filters and change them usually after 40–80 hours of use; check all lines for kinks or obstructions

❏ Clean valves in replenishment pumps for free, unobstructed flow

❏ Store unexposed film between 68–70°F (ambient temperature) in a closed room; in poorly kept storage, emulsion can change speed, affecting box-to-box exposure times and processing times

❏ Monitor film D_{min} and D_{max} throughout the day using a calibrated densitometer

Automatic Silver Recovery Units

❏ Dispose of silver properly by a collection service

❏ Carefully remove silver (many devices prompt you when it's ready; you may also use a silver test to determine silver content) by using a putty knife and following the operator's manual

❏ Wipe clean the drum inside the unit with a cloth anytime silver is removed; remove any fungus or tar that would prevent desilvering from occurring

❏ Clean the exterior with a moist cloth and a mild soap

❏ Lubricate gears once a year only as recommended

❏ Report any non-functioning silver recovery units to management

Analog Proof Processors

❏ Clean entrance feed table and feed roller

❏ Drain and clean developer racks and tanks

❏ Brush pressure adjustment

❏ Lubricate drive-shaft worm gears

Analog Proof Laminators

❏ Clean and lubricate where applicable and specified by manufacturer

❏ Dust laminator daily (including at entry table)

❏ Clean rollers before proofs jam or if dirt or staining is deposited on proof

❏ Clean laminating rollers when a wrap occurs; follow manufacturer's instructions when wrapping proofs become a problem

❏ Clean temperature sensors regularly; filthy sensors cause incorrect reading of actual roller temperature

❏ Replace guide plate cover sheets when soiled or damaged; use only specified heat-resistant adhesive tape when replacing cover sheets

Digital proofing systems, like the large-format Encad NovaJet (shown here) are gaining popularity. Follow manufacturer's recommended specifications for maintenance.

Platemaking

- ❏ Keep plateroom clean; maintain positive pressure (forcing air out, not in)
- ❏ Temperature should be about 73°F, relative humidity 50%
- ❏ Expose step wedges on every plate as QC evidence; processors and frames should be well-maintained
- ❏ Store plates in a cool, dry place, not exposed to light. Oldest plates should be used first
- ❏ Store chemicals at room temperature

Platesetters

- ❏ Have appropriate exhaust
- ❏ Keep lids closed when not in use; assure all covers are firmly in place
- ❏ Check vacuum gauges for adequate vacuum
- ❏ Clean the air filter on the device with a vacuum cleaner
- ❏ Use a vacuum cleaner to remove particles from the drum and around equipment

Plate Processors

- ❏ Clean regularly, fill with fresh chemistry and maintain it according to manufacturer's recommendations; all spray bars/tubes, brushes and rollers should be cleaned frequently and replaced as needed (their pressure should be regularly checked). Use water and fresh developer over rollers to clean them if needed.
- ❏ Check and change filters frequently according to manufacturer's recommendation; "filters are filters" is a misconception, so use the recommended one for each machine

Examples of external drum (Optronics, *right*) and internal drum (Krause, *bottom*) plate-setters.

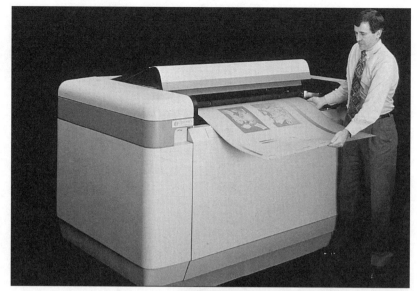

☐ Lubricate according to manufacturer's recommendations
☐ Maintain temperature and strength of chemicals and dryer in the processor according to manufacturer's recommendations; use bi-metal thermometers to verify temperatures. Speed of processor must match media and chemistry used.
☐ Check replenishment rates on a daily basis
☐ Check roller tensions: run different thicknesses of plates and evaluate performance and results
☐ Check guide roller and gearbox for wear or excess play
☐ Check drive shaft coupling alignment with roller couplings
☐ Check chain alignment, tension
☐ Check fixings and fasteners wherever accessible
☐ Maintain proper pressure adjustment of brushes to achieve good cleanout for certain plates and to provide even pressure along the entire length of the brushes. Use a spring scale to ensure proper adjustment.

❏ Store unexposed plates in a cool, dry place not exposed to light. Oldest plates should be used first. Handle plates under safelight recommendations from the manufacturer. Chemicals should be stored at room temperature.

❏ Maintain temperature and strength of chemicals in the processor according to manufacturer's recommendations. Speed of the processor must be correct for the media and chemistry used.

❏ Check replenishment rates on a daily basis

❏ When calling for technical service help for plate processor problems, have the following information handy:
- The plate and chemistry used; evidence of the plate (do not trash)
- Development time (actual, not what the dial reads)
- Development temperature (actual, not what the dial reads)—use a bimetal thermometer for checking all chemistry temperatures
- Gum temperature (actual, not what the dial reads)
- Dryer temperature (actual, not what the dial reads)
- Control strip values if used
- Make, model, and size of processor being used
- Printed sheet showing problem
- Lot number of plates, e.g., 61450000

Imagesetters

❏ Have appropriate exhaust

❏ Keep lids closed when not in use; assure all covers are firmly in place

❏ Clean the take-up and supply compartments with a vacuum cleaner once a month

❏ Open the take-up cassette so felt can be cleaned (once a month)

❏ Clean the air filter on the device with a vacuum cleaner

❏ Clean the outer surfaces of the machine with a damp sponge (not sopping wet) to reduce dust and dirt; do not allow liquids to enter inside imagesetter

❏ Avoid material jam by keeping track of how much material is inside take-up cassettes

❏ If accessible, inspect imaging surfaces for defects which would interfere with image quality

Computers

❏ Turn computers off when doing any cleaning

❏ Clean the computer case by wiping down the surfaces with water-dampened cloth

❐ Clean the monitor by using a glass cleaner, sprayed onto soft cloth, and wipe over monitor

❐ Check settings that affect color viewing such as brightness, contrast and convergence

❐ Calibrate the monitor daily for soft-proofing applications

❐ Clean the mouse by disengaging the ball, brushing out any debris, and replacing ball

❐ Clean keyboard by using an air blower (to evacuate trapped dirt between keys); never use a sopping damp cloth to clean keys (drips down into keyboard)

❐ Vacuum areas around computers periodically so dust on desks doesn't get into computer

❐ Protect computers and all components from sunlight (heat) and moisture

❐ Regularly examine the condition of the system files and directories on your disks with a disk utility. Defragmenting and optimizing disks help them run faster. Deleting, moving, copying, and adding files to hard drives causes fragmentation, but by defragmenting, data that was scattered in different locations becomes contiguous. This speeds up file access and increases reliability. One effective, reliable optimizing program should be purchased.

❐ For Macintoshes, rebuild the desktop once a month. This reduces the size of the desktop files, creates more free space on the disk, and speeds up disk mounting.

❐ For Macintoshes, clear the Parameter RAM by rebooting with the Command+Shift+P+R keys, or by using a utility

❐ Remove the case cover and vacuum any dust inside the cabinet

❐ Check connections of network and peripheral cables

❐ Watch computer magazines and on-line services for the latest system software upgrades and install the latest version on your machine

❐ Virus control is vital in preflighting when accepting clients' files from many sources. Prevent passing a virus to your computers by installing virus detection software which will help to spot viruses before they enter your system and cause havoc.

❐ Keep magnets or magnetically charged objects away from computers and disks

❐ Repair a hard or floppy disk using reliable "Disk First Aid" software

❐ Allow plenty of room around computers for ventilation

❐ Keep liquids away from keyboard in case of possible spillage; disconnect power to computer if it should happen

Floppy Disks
❐ Clean the read/write heads of the drive with an approved cleaner disk
❐ Store all disks where they will not be exposed to excessive cold, heat or moisture

CD-ROM Drives
❐ Use an approved cleaner to clean the optical CD-ROM lens, prevent drive reader error, and improve access to data

Summary

The performance of your production process depends on the integrity of the files you work with. Maintaining the equipment used to produce printable files will help you optimize the entire production process.

8　Optimizing the Press

Producing quality color printing can be a major challenge to many printers. The customer's demands on high quality, consistent quality, and quick production turnaround have forced many printers to rely on make-it-work production control. Make-it-work production control arises from inconsistent scheduling, complex materials and chemical interactions, lack of effective press operation and maintenance, lack of reliable measurement data, and relying on subjective judgment for attaining and maintaining color on press.

The establishment of a printing system with a high degree of predictability requires a structured protocol of process testing and equipment optimization. The foundation for successfully bringing the system to a higher level of quality and predictability is discipline. Discipline requires the printer to develop structured planning and adhere to established standards in preventive equipment and process maintenance, surpassing necessary equipment conditions, consistent sources of materials, and realistic monitoring of production conditions.

Press optimization is fast becoming a requirement, if the printer wants to be competitive and survive. The sequence for optimizing the printing system must begin with the press and work back through the prepress system to color separation of films. The printer must know what mechanical problems may exist on the press. The materials should be evaluated for quality and compatibility. The press's maximum ink densities and dot gain characteristics are then determined. Target ink densities and dot gains should then be calculated and established. Finally, the press data can be plugged into the prepress imaging systems for accurate films, proofs, and plates. Test targets and forms will be required to enable the optimization process to be conducted.

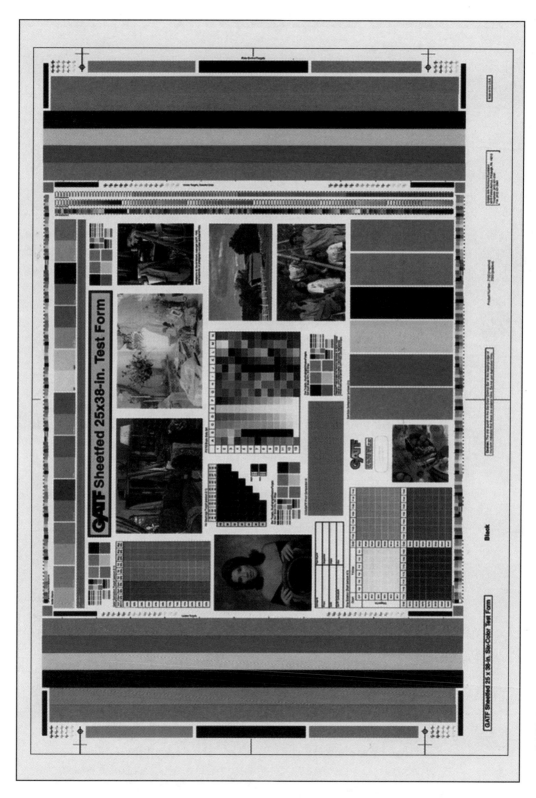

GATF Sheetfed Press Test Form (order no. 7097/7197).

The targets include the GATF Press Test Form plates, the GATF Mechanical Ghosting Form plates, the GATF Register Grid plates, and wet solid plates for both dry and wet solid tests. This chapter is a breakdown of the six steps in the optimization process.

Materials Procurement and Testing

The best possible print materials should be used on the press during the press optimization. The press test is designed to determine what the press is capable of producing. The materials should then be chosen for their quality performance attributes and interactions with the other materials. The important materials that should be tested are plates, blankets, paper/board, ink, and fountain solution. The materials should be tested to determine if they will meet specified requirements for performance compatibility.

Plates

Plates should be tested to first determine image resolution or their ability to reproduce a sharp halftone dot. Second, a register grid test of the vacuum frame or step-and-repeat equipment should be conducted to determine if accepted register accuracy is being achieved. Most lithographic plates are capable of reproducing 8-micron line resolution. The plate's resolution is determined by exposing a plate control target on the plate to the manufacturer's recommended exposure step. Register accuracy should be within ± 0.0005 in., or within one-quarter of a 2% dot.

Blankets

A buffed quick-release blanket should be used for the sharpest possible dot reproduction on press. Although a lower priced everyday blanket may be used to fingerprint the press, the goal is to optimize press capabilities. The better blankets are more stable when torque is applied around the cylinder, and they experience less compression sink and gauge loss. A blanket swell test should be conducted with the printer's solvents. If the rubber face of the blanket swells more than 0.002 in. after four hours of contact with the solvent, then the solvent is too harsh for that blanket.

Paper/Board

The paper or board to be run during the press test should be top quality stock. Recycled stock could be run and evaluated as a second part of the test. Poor quality or old stock should not be used for the test; problems could be created from stock related abnormalities instead of press-related ones. Certain testing should be conducted on the stock to determine acceptance for the press test. The criteria for testing should include:

- Stock coating absorptivity and paper mottle with K&N ink tests
- Surface pick testing with IGT printability tests, or Dennison pick wax test at printer
- Coating surface energy with Dyne level test fluids
- Loose trash on sheet with LandSco light

The type and weight of the stock is dependent on the type of work the printer does. Commercial printers should run on a light- to medium-weight, gloss-coated stock, such as 80-lb. gloss enamel. Label packaging printers should run the press test analysis on a 60-lb. C1S litho stock. Folding carton packaging converters should run their test run on natural fiber stock such as SBS or SUS. If recycled stock is commonly run by the folding carton converter, clay-coated news back can be run as a second part of the test. Remember, the dot gains on recycled CCNB are usually lower than natural fiber stock. Folding carton stock used to run the test should be a mid-range folding carton stock, 16- to 22-pt. caliper.

Ink

The best possible ink should be used to optimize press capabilities. The ink should be capable of producing acceptable dot gain and overprint traps at reasonable ink film thickness, and should match the specified color.

The ink tack, or the ink's ability to overprint-trap over another ink, should be in a two-point spread-rated sequence. Four-color process inks are designed and normally run with black on press unit #1, cyan on press unit #2, magenta on press unit #3, and yellow on the last press unit. The black and cyan are many times at the same tack rating because poor traps of these colors are not normally visible. Following is an example from inkometer measurements for sheetfed process inks:

- Black   Tack 16—1200 rpm
- Cyan   Tack 16—1200 rpm
- Magenta   Tack 14—1200 rpm
- Yellow   Tack 12—1200 rpm

Ink viscosity is another performance indicator. If ink viscosity numbers, from a viscometer measurement, are below 200 for sheetfed inks, higher dot gain can frequently occur.

Water pickup, or how much water will emulsify into inks, can help determine how consistent an ink can hold its density on press. All lithographic inks are designed to take in certain amounts of water or they will not run very well on press. Ink/water emulsification testing results of quality lithographic inks should range between 40–60% water picked up by the ink. Inks with a water pickup of over 80% may make it difficult to maintain ink density on press. Inks below 40% water pickup may have a short flow, resulting in poor ink transfer and piling on the plates and blankets.

Fountain Solution

Fountain solution that is not compatible with the inks or plates can cause higher dot gain. The higher dot gain could be mistaken for ink- or press-related causes. The plate and ink suppliers should be consulted when determining fountain solution compatibility. Typically, a mixture with a minimum of two ounces of etch mixed per gallon of water should be a starting point when established the right concentration. The conductivity of the fountain solution should be established with the help of the fountain solution manufacturer's help. Depending on the manufacturer, fountain solution conductivity frequently measures 1500–1800 micromhos above the conductivity of the water on a calibrated conductivity meter.

Press Specifications and Diagnostics Phase

It must be determined if the various press operation components have been set and maintained to manufacturer and materials specifications. A series of checks utilizing specific test forms should be conducted to verify if the press components are set up and operating correctly. If abnormalities are discovered, corrective action should be taken before proceeding with the next phases of the press optimization process. The series of checks and tests include:

Press Setting and Specification Checks

• Ink and dampening rollers conditions, and settings according to roller and press manufacturers' specifications. The roller shore hardness should be measured with a Type A durometer gauge. New roller shore hardness should be based on press manufacturer's specifications. The common shore hardness on new rollers for many presses with continuous-flow dampening systems are: dampening form rollers 20–25; dampening meter rollers 18–22; ink form rollers 22–27; ink distribution and ductor rollers 25–30. If the shore hardness of the ink system rollers becomes more than 10 points above the new roller specifications, the rollers should be replaced before the press test is run. The proper roller settings and techniques are extremely important to minimize roller streaking problems. The press manufacturer's specifications should be followed for correct stripe settings. Ink-form roller settings will vary depending on the size of the press. For example, 40-in. press ink forms are usually ⅛–³⁄₁₆ inches (3–4 mm) wide across the plate. Larger presses will normally require wider stripe settings.

- Determine if press bearer rings are in good condition and making proper contact when the press is on impression. This can be done with an inked thumbprint technique to verify bearer contact. Use a thin ink film on the thumb, and place it on both plate-cylinder bearers at six-inch intervals. Run four to five sheets through the press on impression. Check for consistent ink transfer onto the blanket cylinder bearer rings. On web presses, bearer-ring contact can be checked using foil strips and measuring the stripe width made on each foil strip. Presses that don't run on bearers will need to be checked with feeler gauges. The press manufacturer should then be consulted.
- Plate-to-blanket squeeze should be set according to the type of blankets being used: conventional blankets (0.002–0.004 in.); compressible blankets (0.004–0.006 in.)
- Plates and blankets should be properly packed to bearer-ring height according to the press manufacturer's specifications.
- Press infeed, transfer register system, and delivery should receive proper cleaning, lubrication, and proper setting based on press manufacturer's specifications.
- Dry solids should be printed on each unit, using process cyan ink on all units, to evaluate the extent of cylinder pressure streaks across and around the cylinder. Cyan ink density across the sheet and from gripper to tail edge of the sheet should be based on a specification, such as SWOP, GRACoL, or PROP. On a Status-T response densitometer, measurement of the cyan ink is 1.30–1.35 in most cases.
- Dry solid impression-cylinder breakaways should be printed on each unit, to evaluate proper print pressure, impression cylinder parallelism, and impression cylinder abnormalities. The impression cylinder is backed away at 0.002-in. intervals, and 25 sheets are run at each interval. Keep running these breakaways until the sheet is just slapping against the blanket, or final breakaway. If different units show printing breaking away at greatly different

pressures, the impression-cylinder gauge settings are probably wrong. The gauges should be reset to proper calibration to prevent excessive impression-cylinder pressure from occurring.

- Print wet solids on each unit to evaluate dampening-system capabilities and roller streaking. If new streaks appear on the wet solids, the dampening system is usually the cause. There could be inadequate dampening-roller settings, poor roller conditions, and incompatible fountain solution concentration.

After the press settings and specification test is completed, it must be determined if any corrective action should be made to the mechanical parts of the press. If so, corrective actions should be conducted before going onto the diagnostic phase. If the press settings and specifications conditions are adequate, the press diagnostic run can now be performed.

Press Test Run and Diagnostic Study

The diagnostic press test is a three-step test process. The first step is to determine mechanical ink ghosting inadequacies, using the GATF Mechanical Ink Ghosting Form. Next, a print-unit register test is conducted using the GATF Register Grid films and plates. The final step is running the GATF Sheetfed or Web Test Form to determine register and printing capabilities. A series of tools is required for the analysis of the diagnostic press test run. The tools include a calibrated Status-T densitometer, and 25× and 100× lighted magnifiers.

Mechanical Ghosting Test

The GATF Mechanical Ink Ghosting Form (order no. 7074/7174) should be run on each unit. Normally, PMS 477 brown ink is run. This ink is used to determine the true mechanical ink ghosting attributes of the press printing units. If the ghosting form reveals ink density differences of more than 0.08 at the intersection of the longest vertical bar and the horizontal bar, measured with a Status-T densitometer, some type of mechanical corrective action may be needed. A change in the type of ink-form-roller rubber may be needed; oscillating ink forms may be another answer, or the ink system vibration drum rollers may require resetting.

Press Units Register Test

The GATF Register Grid (order no. 7160) with original color control bars burned in should be properly mounted on the print units in accordance with the press's pin register system. Each unit must be a different color ink: black, cyan, magenta, yellow, Pantone warm red, and Pantone green. Run up to even ink densities across the sheet and print a minimum of fifty sheets. The first pull register should be within one grid line at the gripper edge of the sheet. If the grid lines exceed the tolerance, the pin register system, plate-clamp integrity, and plate-cylinder zero set should be evaluated for nonconformity and corrective actions taken. The plate and/or cylinder moves should be performed to register and fit the lead edges. There are no image-fit accuracy standards for presses. Half-size sheetfed presses (25–28 in.) can achieve dot-for-dot register at the lead edge and up to one half of a 2% dot at the tail edge of the sheet. Larger 40-in. presses normally can register to within one-and-a-half 2% dots at the tail. Web presses can reveal closer register fit at the tail. An unusual sound will come from the press when running the grid test. This is due to the line repeats on the grid. If the lack of vertical fit at the tail edge of the grid is excessive, the cylinder circumference may not be consistent on all cylinders. A plate and blanket packing change may be required to bring the register and fit into acceptable tolerances.

Test Form Run

The GATF Press Test Form plates should be installed on press for the actual press test run. This test form includes an array of quality targets that enable the test form to perform press diagnostics, process capabilities studies, process control, and standardization.

After the plates have been inspected, the makeready process begins with mounting the plates on the press. Matching register and fit specifications of all colors begins with the first pull sheets. An examination of the first pull sheets is conducted to determine pin-register accuracy. Register and color fit are accomplished with the aid of the transfer grids on the test form. The inking system is adjusted to achieve a production ink profile. The initial ink-density profiles will be adjusted to achieve ink/water balance and run the diagnostic phase of the test run. The initial ink profiles can be based on the shop's typical ink densities or SWOP ink references, available from the International Prepress Association. The color control bar across the tail of the test form is used to achieve consistent ink-film densities across the sheet within ± 0.05 on a status-T densitometer.

The press speed (impressions per hour) of the test should be between 80–85% of the press's rated speed. The press should be allowed to run at the test speed for 500–750 sheets to

assure that ink/water balance is achieved at the target densities of all colors across the test sheets. To complete the test run
the press should continue to run another 1000–1500 sheets.

The diagnostic targets are analyzed after the completion of
the run and the ink has had a chance to complete the early
drying stages, in order to determine if any press printing
problems exist. The GATF test form targets include the ladder targets, star targets, Dot Gain Scale II targets, vernier
targets, and mottle patches.

Ladder targets (GATF order no. 7090/7190) are located on
each side of the test form around the cylinder from the lead
edge to the tail edge of the sheet or signature. The key elements of ladder targets are the vertical and horizontal lines
covering 50% of the area. For example, vertical movement or
slurring will cause the horizontal lines to gain resulting in a
darker image. The same applies for horizontal movement
affecting the vertical lines. Ladder targets will reveal very sensitive physical printing and mechanical problems. The problems include excessive ink dot gain, poor sheet register infeed
and transfer, poor leveling of web units, lack of ink/water balance, loose blankets, gear deterioration streaks, gear eccentric-
ty, inconsistent cylinder bearer contact, sheet and web fan-out,
and dot transfer doubling and slurring. Measuring with a
status-T densitometer, a density difference between the vertical and horizontal lines at the same location should be ≤0.05
for most colors and 0.01 for blacks and very dark colors.

Star targets (GATF order no. 7004/7104) are positioned at
a number of key points on the test form to give a quick visual
diagnosis of potential excessive dot gain problems. If good
printing conditions exist, the star targets will display a small
round dot at the center of the star target. If problems exist,
star targets can reveal three different printing abnormalities. Excessive ink film or a high-dot-gain ink will display an
abnormally large round dot at the center location. Dot slur
will display a football shaped center dot, perpendicular to the
direction of the slur. If doubling of the dot is revealed, the
star target will visually display a figure eight or bow tie.

Dot Gain Scale II targets (GATF order no. 7051/7151) are
designed to give physical dot gain a number amount. If the
scales reveal physical dot gain of 8% or higher, there is excessive dot gain occurring.

Mottle patches visually reveal the existence of ink mottle.
Ink mottle, uneven ink lay, can occur in both single ink films
or multiple-ink overprints. Mottle from uneven ink absorp-

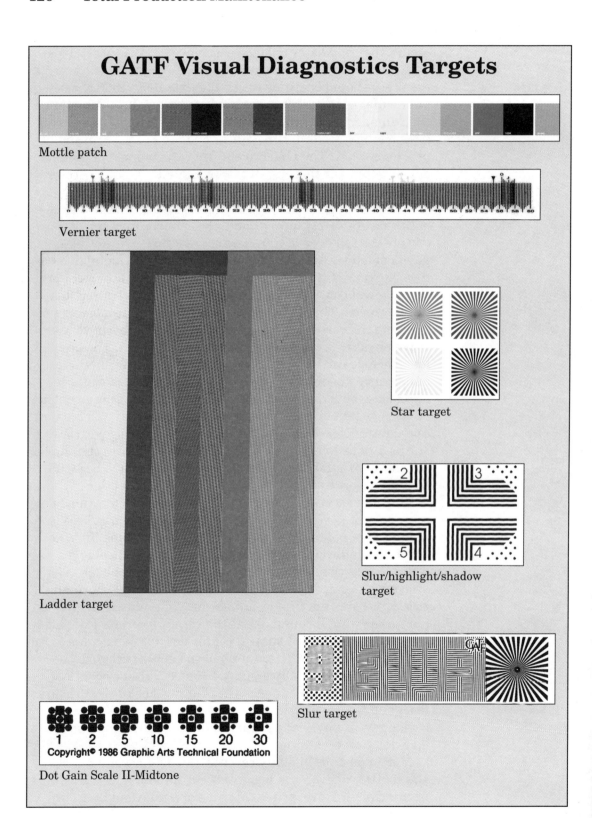

GATF Visual Diagnostics Targets

Mottle patch

Vernier target

Ladder target

Star target

Slur/highlight/shadow target

Slur target

Dot Gain Scale II-Midtone

Copyright© 1986 Graphic Arts Technical Foundation

tivity is normally a paper problem. Mottle of ink overprints is normally caused by an ink compatibility or ink/water balance problem. Ink overprint traps can be measured automatically with many status-T densitometers. Good overprint ink traps normally measure:

- magenta/yellow—red = 76–84%,
- magenta/cyan—blue = 72–78%,
- cyan/yellow—green = 85–92%.

Ink mottle is a more common occurrence in the blue traps, the cyan and magenta overprints.

Vernier target (GATF order no. 7005/7105) measurement and evaluation determines print length differences between the units, and paper stability. Paper dimensional fan-out and print length stretch is measured at both the lead and tail edge locations of the form. Transfer register between print units is measured within eight inches of the sheet's lead edge location. If transfer-register inaccuracy exceeds 0.00075 in. (0.02 mm), a visual contrast in the physical test form targets will normally occur.

If abnormal printing conditions exist, corrective actions to the press should be taken before proceeding to the press capabilities and characterization phases of the optimization process. Another diagnostic press analysis should be conducted to verify the effectiveness of the corrective actions.

The thoroughness of diagnostic analysis is very important, if effective corrective actions are to be accomplished quickly. The GATF Preucil Print Analysis Laboratory has developed three levels of press analysis, one of the most comprehensive press analyses in the industry.

Press Capabilities Study

Once satisfactory printing press conditions have been achieved, the final stages of the printing press optimization process can be carried out. There are two tests that must be done to determine the press's capabilities: first, a test to determine what are the press's highest inking levels or capacities; second, a test to determine printing and inking variability. These two tests will provide data to determine the press system's ink density and dot gain tolerances.

Inking capacity can be determined by systematically increasing the ink density levels after reaching the originally-targeted ink densities. Samples are taken as the ink densities increase, at about 5% ink density increase intervals. The densitometer print-contrast measurement is conducted and calculated on the increased ink density of each sample sheet. The data is input on a X/Y graph, Y axis for print contrast, X axis for increased ink densities. The resulting curve will give the print contrast picture for that press. Print contrast values normally will increase as ink densities are increased. The print contrast values will start to decrease once maximum ink densities are surpassed and dot gains become excessive. Further increases in ink densities will cause print contrast values to drop. The highest point of the curve (the apex) reveals the highest print-contrast value and the highest density that the press can achieve for the ink color evaluated. The highest ink density point should be the upper control or tolerance limit for press production. The maximum inking capacities have now been determined.

The next step is to establish the lower inking control limits or tolerances. To be able to determine the lower inking levels, a variability study of the press's normal ink levels is necessary. The variability study is designed to show how much ink variation occurs during a production run without press operator intervention. Choosing average jobs with medium ink coverage should represent the normal production runs in the plant. For a more effective variability study, choose the normal densities targeted for regular production runs. Once the press has reached ink/water balance the press should run at optimum speed (minimum of 80% of the press's rated speed) for a minimum of 1500 sheets. From the pressrun, at least 100 random printed samples should be collected for the variability study. The 100 samples should be kept in the order they were pulled from the pressrun. Measuring at a specified location on each sheet, the densities are measured with a densitometer and the data statistically analyzed. From the analysis, the mean and standard deviation can be calculated. The mean value should normally fall close to the original density aim points of the variability pressrun. The upper and lower process control limits then should be based on the natural ink density variation or the standard deviation measured from the study. Most manufacturing processes will normally consider ±3 standard deviations as accepted tolerance. The target densities for production printing should then be set at 3 standard deviations below the maximum densities determined from the ink capacity study. The density tolerance limits are now ±3 standard deviations from the newly established target ink densities. Each process ink will have different tolerances. The next thing to examine is the overprint trap percentage values at the new target densities. If the overprint traps have significantly declined, an adjustment to the ink densities levels may be necessary.

- Input each density value on a histogram.
- Evaluate the bell shaped curve created in the histogram, the mean and the standard deviation can be calculated from the data.
- You can then determine where the data bars are situated and the center point of the distribution. The center point will give the most frequent density value during the variability study.
- The width of the bell curve will give the ink density distribution, the plus-or-minus density variation of the press.
- The mean and standard deviation values can now be calculated using a scientific calculator.

Achieving optimal ink densities and overprint trap values are main steps for ink-color control in daily press operations.

Press Characterization

The last phase in the press optimization process is characterization of the press system or the true fingerprint of the press. It is important that the press troubleshooting and capability phases be completed before the press characterization phase begins. Many graphic communications industry suppliers attempt to provide the quick fingerprint of the press for their customers, without troubleshooting the press or determining inking variability and maximum ink densities. That type of press analysis may very well be an inaccurate profile of a press's capabilities. Press characteristics must be established from good or optimum running conditions. The GATF test form contains a number of targets that will enable measuring the press's printing characteristics. Targets include: color control bars, tone scales, dot size comparator, ink coverage target, gray balance charts, two color overprint tone scales, and the GATF Plate Control Target.

From the color control bars and tone scales, dot gains of each ink color are measured through the reproducible range of halftone dots. Inputting the dot gains on a chart, a dot gain curve is constructed. The dot gain curves will enable prediction of the film halftone dot sizes from the apparent printed halftone dot sizes.

The next target, the dot size comparator, will enable measuring printed dot sizes at different screen rulings (133, 175, 200, and 300 lpi). The dot size comparator allows comparing

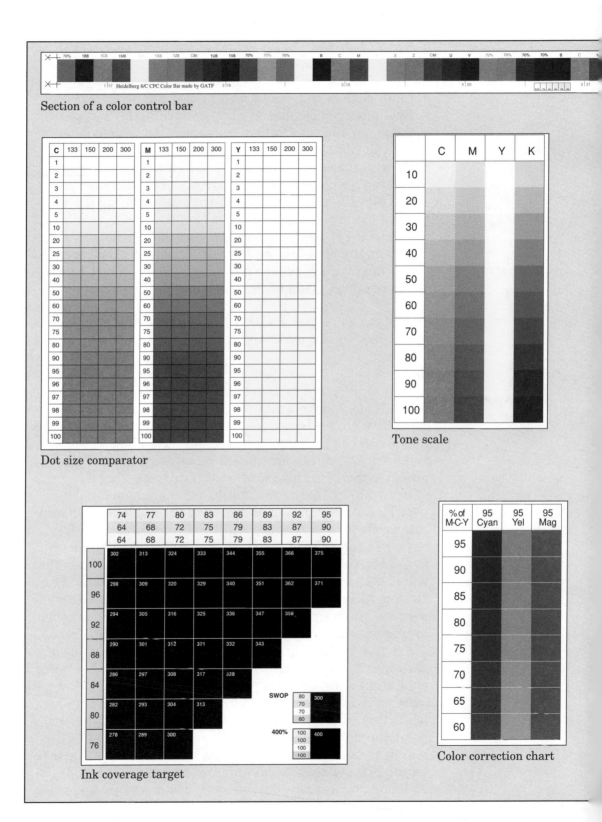

Section of a color control bar

Dot size comparator

Tone scale

Ink coverage target

Color correction chart

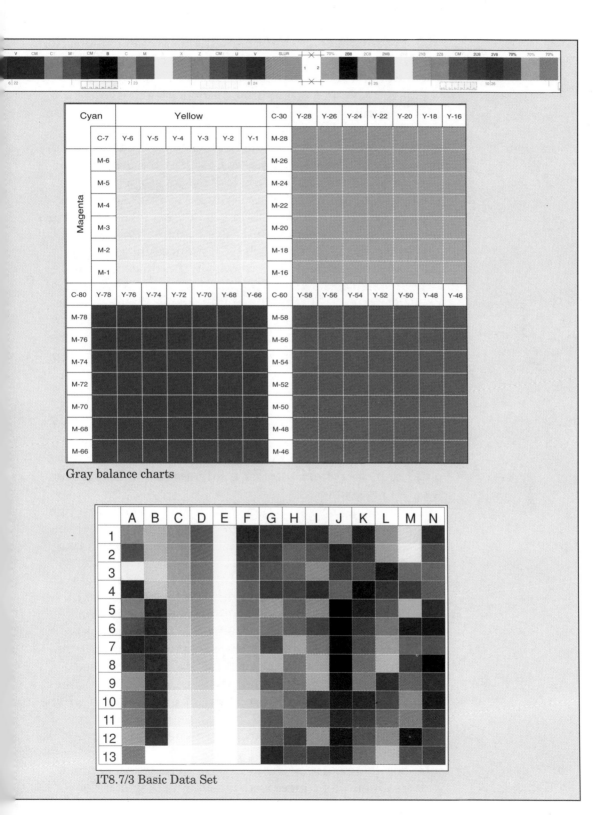

Gray balance charts

IT8.7/3 Basic Data Set

150-lpi dot gain characteristics to higher screen rulings. Normally, dot gain will increase at higher screen rulings, with the lower screen ranges (20–30% dot sizes) increasing more.

The ink coverage target provides densities that are produced from various combinations of black, cyan, magenta, and yellow ink overprint coverage. The target is designed to show the effects on shadow density when reducing from the maximum ink coverage of 400%. Lowering the ink coverage to 360–340% will enable printing less ink coverage in shadow areas and minimizing problems caused by excessive inking. It is important to specify the optimum ink coverage of the printing press system when applying, for example, gray component replacement (GCR). Too much GCR reduced-ink coverage may cause insufficient ink coverage in the shadow areas.

Gray balance charts provide the three-color grays for that printing system at the main points through the tonal scale. The gray balance charts can either be visually compared to a certified gray scale, or measured with a spectrophotometer and integrated with a color measurement software program. Acquiring the desired cyan, magenta, and yellow film dot sizes is needed to produce neutral gray on printed sheets.

Highlight and shadow elements on the GATF Plate Control Target provide the largest shadow dot and the smallest screen dot the printing system will successfully produce. This information will provide the color scanner or the computer-to-plate system with the minimum and maximum dot it needs to produce.

Color hexagon construction enables comparison, on a light table, of printed sheet to proof, and of approved printed sheet to printed production sheet. Measurement of specified two-color overprints of red, green, and blue tonal scales are combined with the cyan, magenta, and yellow tone scales to construct the color hexagon. Multiple color hexagons will provide information for comparing several print system attributes of press to proof and pressrun to pressrun. The hexagon can be plotted from samples taken throughout the pressrun. The increased color saturation from increased screen tints and solid densities will plot on the chart farther from the center point. The process color print contrast will be the distance between the 75% hexagon and the solid process ink coverage points on the hexagon, higher print contrast will result in wider spacing. The 10%, 25%, and 50% process-color dot gain points on the hexagon's spacing will become wider, influenced by increased density on dot gains. The

The GATF Color
Hexagon shows informa-
tion about tone reproduc-
tion, density, and hue
error.

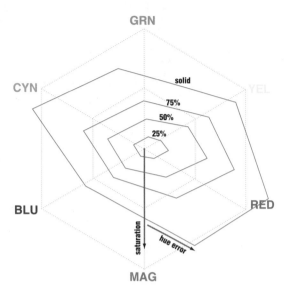

blue, green, and red points for 10%, 25%, 50%, 75%, and
100% may show less spread with increased ink coverage and
lower overprint traps.

A true press fingerprint is the characteristics of a press
system, provided by a certified (GATF) test form, that is
operating at its optimum.

**Press
Calibration
to Prepress
Systems**

The fingerprinted press values must be entered into the prepress systems. The color proof is a visual connection between
the color separations and the pressrun. Choosing or adjusting a color proofing system is necessary to be able to provide
adequate matches on press. The GCA/GATF Proof Comparator target on the test form is the tool that will enable
calibration of proofing. A proof is made from the Proof Comparator films and compared to the Proof Comparator on the
test run sheets. Adjustments are made to the proofing system or a different proof system is chosen to achieve an
adequate match. The halftone dot gain range values, and the
minimum and maximum halftone dots from the test sheet
analysis, can now be entered into the digital prepress and
color separation process. The result should be films that provide predictable dot gains based on that press's optimum
capabilities.

Summary

The GATF Press Color Test Form is a press optimization tool that will enable the printer to diagnose and mechanically troubleshoot the press, establish maximum ink density capabilities, determine ink density variability, and characterize the press's fingerprint of optimum ink densities and dot gains.

9 Quick-Response Makeready

Today, customers are requiring shorter run lengths, tighter lead times, and greater quality from printers. Due to the increasing popularity of just-in-time (JIT) concepts, no one wants to keep large inventories. This means that printers are running more jobs with shorter runs. However, during makeready, the press is not producing salable printed sheets. Thus, the printer must develop more efficient, and more effective, press makereadies to be able to compete in today's printing market.

Press Makeready

What is press makeready and what does it encompass? Press makeready is the transition process between completing one job and preparing the press to print another job. In other words, makeready is what happens between the printing of the last good product sheet of one job and the first good product sheet of the next job on a particular press. That time usually includes the following tasks:
- Wash up and clean up the press
- Change plates
- Check plate-to-blanket squeeze on each unit
- Set up the press feeder and delivery
- Set up the inking system
- Set up the dampening system
- Change the coater blanket (if applicable)
- Make trial impressions for register and color
- Match job specifications
- Get approval to run (OK)

These tasks are normal, traditional requirements during press and pressroom operations. Shorter makeready times will help the press become a greater income-producing center for the printer. Long makeready times will cause the press to become a cost center instead of a profit center.

Single-Minute Exchange of Die

One way to develop effective press makereadies is through a system called single-minute exchange of die (SMED). SMED was developed in Japan by Shigeo Shingo. SMED is a scientific approach to reducing inefficiencies and redundancy in work and setup processes.

SMED does not represent a new way of doing makeready. Rather, it is a scientific approach to traditional press makeready. The difference lies in when and how specific tasks are done. This chapter is not about SMED, except as an approach to cutting makeready time.

To analyze any process, SMED differentiates between "internal" and "external" tasks, which together encompass the setup for a particular machine. Internal makeready tasks can only be performed when the press is stopped. External tasks can be performed when the press is in running operation. The SMED approach to makeready encompasses four key steps:

1. Identify makeready elements
2. Separate internal and external elements
3. Convert internal elements to external elements
4. Streamline both internal and external elements

Identify

To help identify internal and external tasks, checklists can be created to itemize everything necessary to perform makeready. Checklists should name each item used, list press crew members and their job descriptions, include equipment and material specifications, and specify press settings.

Separate

To determine what happens during a makeready, it must be videotaped and carefully analyzed to separate the internal and external tasks. When analyzing the video, the press crew must ask several questions:

- What is the purpose of each step?
- Why must the press be stopped to perform that step?
- How can the step be converted from internal to external?

Technical checks are done to confirm that all materials (paper, inks, plates, blankets, rollers, instruments, gauges, tools, etc.) meet the requirements of the job and are appropriate for the press.

Operational checks are done to confirm that everything required to perform tasks is in place. Function checks are done to confirm that everything works properly.

Convert

The next step is to convert internal tasks to external tasks. In other words, how can tasks currently performed when the press is stopped, be performed when the press is running? Converting as many tasks as possible minimizes press downtime and maximizes press production.

To help you get started, following are some examples of makeready tasks that can be completed while a job is running on press.

- Finish paperwork. Use carbonless forms to reduce redundant documentation.
- Check plates for correct image, voids, and nonimage spots.
- From the remote control console, preset the ink fountain keys for the next job.
- Zero-set the register control systems before stopping the press washup.
- Develop staging areas for materials.
- Develop staging racks for plates, inks, and job envelopes.
- Stage the materials at press.
- Stage the washup components (washup trays, solvent bottles, water bottles, and towels) at press units.
- Zero-set the ink fountain keys to clean off excess ink before starting the press washup. Clean and gum plates for storage.
- Schedule jobs to run on press according to common ink type, ink sequence, and stock size.
- Return materials and tools to appropriate staging locations.
- Clean ink knives and washup trays while the job is running.

Streamline

After converting internal tasks to external ones, both must be improved and streamlined. To do this, it is a good idea to document and analyze each task and simplify, combine, and eliminate whenever possible.

Establish standard operating procedures to maximize the efficiency of every press operator's every move. When the press crew works as a team, several tasks can be done simultaneously.

Improve methods for cleaning ink fountains, rinsing ink rollers, changing plates, and setting up the press feeder and delivery. For example, one press operator devised a better ink scoop by modifying a scoop for a different commercial

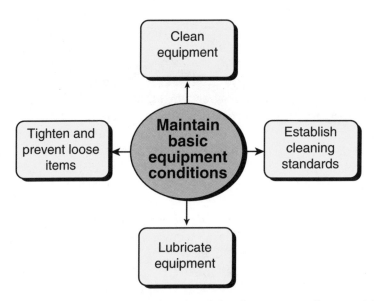

application. He can remove 80% of the fountain volume with just four scoops.

Mechanize tools to improve press operator efficiency. Or, acquire new equipment and tools if needed. At one company, using a pneumatic wrench cut the time in half to remove coater blankets.

In addition to streamlining actual tasks, materials and tools represent another opportunity to trim time from the makeready. By ensuring that the quality of materials is sufficient for the job, many problems can be avoided—problems that are more costly further along in the production process. Building or purchasing transport carts, lifts, and extra staging areas would allow materials to be set up off-line.

Refining maintenance and storage of tools that are used during makeready also saves valuable time. Use common tools to reduce or eliminate the need for internal adjustments. Obtain extra tools. Learn to use new techniques and tools, as well as pin systems, plate scanners, gauges, etc., to make various tasks more accurate and efficient.

Evaluate job jackets. Even though the jacket should contain complete instructions for the job, the jacket may not ask for all the necessary information. In addition, it may need to be redesigned to become more "user friendly." One printer used to process about twenty job jackets per day, and at least four or five would have to be corrected. For each one, the pressroom supervisor had to go to the salesperson or customer service representative (CSR) to have the instructions clarified or completed. After a few months with the new jacket, corrections were down to one or two per week.

- Develop a standard operating procedure on plate mounting, including proper use of the pin system, with the goal of getting the first pull in register.
- Better tools, such as torque wrenches, power wrenches, and powered lifts, can improve and speed required tasks.
- Develop standard operating procedures for press washup. The press crew needs to work as a team and perform various tasks simultaneously.
- Lift trucks can efficiently move materials for staging.
- Develop techniques to ink up the press to lead to quick color approvals.

The Role of Managers

For the SMED system to work, first there must be a total commitment from the printing company's managers and production employees. Top management is responsible for company-wide awareness and support for improvement. Top management must also provide resources, leadership, and total support for middle managers and press operators.

Middle management must provide leadership and support for press operators. They must develop and implement solutions to problems. The middle manager today must manage, facilitate, and coordinate department operations. The sign of a good manager is in how well problems are prevented in a department, not in how well problems are resolved for the short term.

Teamwork

The press operators implement the SMED system in the press department through teamwork. Through press team meetings, the press operators evaluate the makeready videos, help identify problem areas affecting press operations, and suggest possible solutions. They must also improve the procedures for both press makeready and operation. Most importantly, the press teams must identify all tasks currently being done in the internal time frame that can be done in the external time frame.

A key factor in the SMED system is team participation from all employees in the company. Technical training and updates are important factors in developing teamwork. The company must see to it that managers and press operators know and understand the technical process they are working on. In other words, they must know the basics and not drift away from them. Press operators must also be given the opportunity to continually learn new developments in technology. Then this increased knowledge can be applied to current press operations as part of the continual improvement process.

The responsibilities of each member of the press crew must be established and spelled out so that the crew can work together as a team. Everyone in the press crew must participate in the makeready process, performing tasks simultaneously.

Process Optimization

Management is also responsible for bringing the process up to its optimum capabilities. The platemaking department must supply plates that are accurately punched and in register for proper fit. They must also be correctly exposed and developed to minimize plate-related dot gain or dot sharpening. Then the plates should be thoroughly inspected for correct image, correct position, broken type, voids in image areas, spots in nonimage areas, etc. After inspection, the correct identification (job number and ink color) is marked on the clamp area of the plate.

Developing a clear, concise, and thorough job information envelope or docket is very important. The envelope must include all required information, with no changes scribbled in.

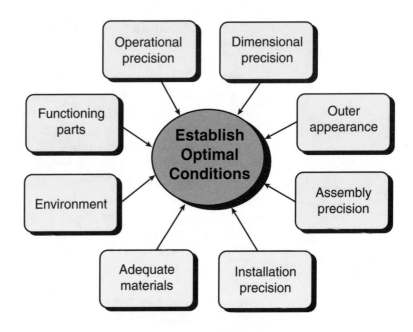

The press materials must be technically correct, compatible, and consistent. These include the stock, ink and coatings, dampening solution, blankets, rollers, etc. These materials should be tested regularly to determine both consistency and compatibility.

The press must be checked to determine if the adjustments and settings meet the manufacturer's specifications. The press must also be diagnosed to detect any mechanical problems. If the press is not set at the manufacturer's specifications, then it must be readjusted to correct settings for plate-to-blanket squeeze, roller pressure, etc. Any mechanical problems must be corrected. Maintaining the press equipment at manufacturer's and supplier's specifications will help minimize unscheduled downtime, ensure good print quality and make problem solving quicker and easier.

When the press is operating at its optimum, it can be "fingerprinted." Fingerprinting a press will give the printer the printing characteristics of the press. The GATF Sheetfed Test Form (see Chapter 8) is designed specifically for this purpose. The test form helps printers to diagnose potential mechanical problems; establish standards for accurate color separations and proofing; and determine run control capabilities of the press through densities, dot gain, and overprint traps throughout the pressrun. Knowing the capabilities of the press will help in standardizing the process.

Pre-makeready

External Tasks

Pre-makeready encompasses many areas that seem unrelated to the press makeready process. Pre-makeready must have a solid foundation or makeready will be disjointed, inconsistent, and many times ineffective. Pre-makeready includes:

- Sales
- Customer service
- Production control (scheduling)
- Prepress production
- Pressroom maintenance and procedures
- Material purchasing and staging
- Quality assurance

These functions are considered "external" because they all can and must be performed correctly prior to the pressrun for that particular job. (However, they can be done while the press is running a different job.)

Communication and Planning

Good communication with external as well as internal customers (production department) is essential for success. Open, company-wide communication will greatly help to overcome problems related to customer changes, scheduling conflicts, exceeding customer expectations, improper material purchasing, etc. Company-wide communications will help in developing a job planning procedure, which will help minimize unforeseen problems in all production departments and ensure that the production process goes smoothly.

Production planning meetings that include customer service representatives, production schedulers, and people from the prepress, press, and finishing departments must be held to review requirements for each printing job, which include:
- Layout (sheetwise, work and turn/tumble, or perfecting)
- Film (positives, negatives, dupes, color separations)
- Type and size of plates
- The press to run the job
- How many colors, type of ink and ink coverage
- Coating (varnish, aqueous based, or ultraviolet)
- Type and size of stock (coated or uncoated)
- Job run length (sheets required)
- Finishing (trimming, folding, diecutting, scoring, etc.)
- Any out-of-the-ordinary requirements
- Delivery date

All questions concerning the job must be asked and answered during the planning stage. All guesswork must be eliminated before the job enters the workflow process. Pre-makeready is divided into two parts:
- **Pre-makeready 1**—Top management, middle management, and staff are responsible for ensuring that all required makeready information, materials, tooling, and equipment are press-ready and at designated press locations before the last job running on press is done.
- **Pre-makeready 2**—Equipment team operators and middle management are responsible for ensuring that all required makeready information, materials, and tooling are press-ready and staged at designated locations on the press, before the last job is done.

If Pre-makeready 1 is not completed, Pre-makeready 2 can not be done. Quick makeready can only be achieved when Pre-makeready 1 and 2 are consistently accomplished.

The salesperson is not only responsible for selling the job to be printed, but also seeing that the job starts off on the right foot. The customer's expectations must be conveyed to the CSR. The salesperson may also need to see that the customer supplies the correct job information and materials (film, proofs, color separations) in a timely manner.

The customer service representative acts as a liaison between the customer and production. The CSR must coordinate the job throughout the process, including the job planning concerns above. It also includes making sure that the information and materials supplied by the customer are complete and acceptable; that all the correct job information is documented on the job envelope; that the stock is ordered; and that the job is scheduled for production and delivered to the customer on time.

The production scheduler must be kept in constant communication with all departments in the production process. Jobs should not only be scheduled according to their delivery date, but also, if possible, according to ink sequences, job layout, and stock type. Good scheduling can often gain valuable press time by minimizing press washups and unit color washes. The production scheduler must also recognize the need to schedule equipment maintenance, which is easier to deal with than unscheduled maintenance due to breakdowns. An effective way of tracking production jobs is to use a computer data input station, interfaced to a central computer in all production departments.

The prepress department's main responsibility is to provide plates for the press. The plates must be properly exposed and developed. It is up to the prepress department to make sure that it receives the correct information and the proper materials to produce plates. Inconsistent platemaking quality will cause costly downtime in both prepress and press departments if remaking plates is necessary.

The pressroom must be supplied with materials that are compatible with each other and of acceptable quality, in addition to the job information, approved proofs, and plates. These materials include paper, ink, blankets, dampening solution, rollers, press washes, plate storage gum, and tools.

Test procedures to determine quality, consistency, and compatibility should be developed. Working closely with the suppliers of these products will help in developing procedures and minimize material-related problems.

The pressroom must develop proper maintenance and cleaning procedures. If the press is properly maintained according to schedule, there is less likelihood that press-related problems will cause unscheduled downtime. The press lubrication schedule recommended by the manufacturer should be followed strictly. Daily and weekly cleaning should be done of the feeder mechanism, sheet transfer system, sheet delivery system, and printing units. Roller maintenance should be scheduled regularly to check roller conditions and settings.

All materials required to run the job on press must be checked and staged beside the press. The press operator cannot afford to walk around the production areas looking for materials. If the job cannot be staged, then it is not ready for press; it should not be put into the production schedule.

During the pressrun (external time), the press operators must prepare for the end of that pressrun and for the make-ready for the next job (internal time). The washup components must be ready and in place: washup trays, full press wash and water bottles, and clean rags. All the materials needed to print the next job must be at press and ready to go.

Press Makeready

Internal Tasks

During the pressrun, the next job should be staged at the press, and all washup components should be ready. Once the last good product sheet has been printed, the actual make-ready begins for the next job. It is a good idea to record settings and adjustments made to the press so that, in case

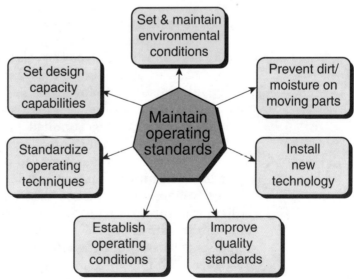

a reprint is ordered (or a similar job is to be run), the make-ready can be conducted faster.

Washup and Cleanup

To scrub blankets clean, use clean towels and blanket cleaning pads with a proper blanket wash and water. Automatic blanket washers may not thoroughly clean the blankets after a production run; coating buildup may necessitate hand cleaning of the blankets. The same procedure can be used to clean the impression cylinder.

Properly gum up plates for storage, if storage is required.

To remove ink from the fountains, run the ink down to a low level (the level depends on the amount of ink coverage on the last job). Ink fountain keys should be zero-set to help clean off excessive ink from the ink fountain ball roller (this can be done when using a remote ink control console). Many press manufacturers recommend using plastic ink knives to remove ink.

Ink fountains must be cleaned with clean towels and proper roller wash. Three wet press wash rags and one dry rag are usually enough to clean out the ink fountain. The ink fountain must be thoroughly clean before adding the next ink to the fountain. The area around the ink fountain must also be wiped clean of any ink and press wash.

Mount the washup trays; make sure the blades are clean and undamaged.

Rinse rollers with roller wash and clean water, using separate application bottles. A two-step roller wash is recommended. Applying excessive amounts of roller wash and water will splash ink and roller wash on the other areas of the press, causing a mess and contaminating the fountain solution. Additional time will be required to clean up this unnecessary mess. If more than a few minutes is required to rinse the rollers, then a problem may exist. The rollers and washup blades must be checked for damage and for the correct setting. The last step of every roller rinse should be to apply clean water to flush off any water-soluble contaminants and chemistry, which are not removed by the roller wash.

Units that are going to print a lighter color ink should be color-washed at this time. Apply either the ink to be run next or a special cleaning compound and let it run into the rollers a few minutes. Rinse off with roller wash and water. This sequence should be done as part of the initial washup, while the dirty washup trays are still on the press. Ink knives can be cleaned at this stage.

Dampening (metering) rollers must be cleaned with an appropriate dampening cleaner. Proper cleaning of the dampening rollers is essential for good fountain solution control on the plates.

While the rollers are being rinsed, one of the press crew members can put ink to be run on the next job into the ink fountain. This will save valuable time during the makeready.

To finish the cleanup, remove the washup trays and clean up the area. Cleaning the washup trays and returning washup items to assigned areas should be done during external time (when the press is up and running that job).

Changing Plates   Carefully remove the plates to avoid damaging them. Plates should be cleaned, gummed, and placed in temporary storage until after the job has been accepted by the customer.

Clean the plate cylinder. Plate clamps must be properly set and cleaned (if necessary) before the new plates are mounted on the press. The gripper edge clamps must be set in the correct position according to the pin system being used. The tail clamp should only be backed away enough so that the tail edge of the plate will insert into the clamp. (Backing away the clamps further than necessary is a waste of valuable time.) The gripper edge clamp, when locked onto the plate, should have even pressure to prevent any distortion to the metal plate, which can cause image fit problems.

Place the right thickness of packing sheets on the back of the plate and roll the cylinders around on impression. Permanent packing is a highly recommended substitute for paper packing. Permanent packing is a plastic sheet the same thickness as paper packing. However, the plastic sheet is attached to the plate cylinder by an adhesive backing. Proper care of the permanent packing will enable it to stay on the cylinder for long periods of time. Using permanent packing is another way to save time during the press make-ready process. The tail edge of the plate can now be inserted into the clamps and locked onto the plate. Now the plate can be snugged against the cylinder by torquing the clamp bolts evenly. A plate-mounting procedure should be developed for each press and pin system. The plates must be mounted the same way every time to ensure consistency.

After the plates are mounted, check the plate-to-blanket squeeze using a calibrated packing gauge. The plate, with packing, must equal the manufacturer's specifications to the height of the cylinder bearer ring. The blanket height (with packing) to the cylinder bearer ring must give a plate-to-blanket squeeze of 0.002–0.004 in. for conventional blankets or 0.004–0.006 in. for compressible blankets. Too much or too little plate-to-blanket pressure will result in print quality problems.

Set Feeder and Delivery

The stock must be reasonably flat and accurately trimmed for consistent size and squareness. The stock pile must be winded and evenly piled and lined up in the feeder. The front stops and side guide must be lined up and working properly based on centering of the sheet dimensions. The delivery must be able to evenly stack the sheets into the delivery pile. Otherwise, sheets may jam in the delivery section causing frequent press stops and ink setoff in the pile.

If the feeder is not set up properly, the press may have difficulty feeding sheets into the press. Poor sheet control results in sheet-to-sheet misregister, frequent press stops (which cause color variation and ink/water imbalance), plate problems and excessive lost press time and paper waste.

Ink System Setup

Clean ink should be used on all jobs. Check previously opened cans of ink to determine if any contaminated ink has been replaced in the can. If so, the ink should not be used.

Before placing ink in the fountain, check to see if the fountain is properly locked into position. The proper amount of ink should be placed into the fountain, depending on the ink coverage and run length.

If the press has a remote control console, the ink keys can be set according to the amount and position of the image on the plates, prior to mounting the plates. Presses with remote control consoles can usually program a special inking mode designed to put even amounts of ink on the rollers, depending on the coverage. This will allow one press operator to apply ink to the rollers very quickly, even on a six-color press. Once the inking is complete, then the fountains are reset to the job programmed in earlier. If the press does not have a console, ink keys can only be set after the plates are mounted and ink is placed in the fountains.

Dampening System Setup

The dampening system must be in top form, or else poor print quality and excessive makeready and production times will result. Continuous dampening systems should be checked to determine whether they have been reset to the printing mode and properly cleaned. Dampening rollers must be set with the correct contact between metering, chrome, and form rollers. Rollers should also be checked for any damage or low areas. The dampening speed control indicators must be checked on all units.

Coater Blanket Change

Setting the coater properly helps minimize cleanup after the pressrun. If an overall coating is to be applied to the sheets, cutting only the outside edge of the blanket or packing is required. This can be done off press (during external time) based on the center point and stock size. The coating blanket should be properly snugged against the cylinder with a torque wrench to ensure consistent transfer pressure and register. Poorly mounted blankets can result in coating buildup in nonimage areas that will contaminate other cylinders and gripper bars with coating, resulting in a lengthy cleanup.

If spot coating is required (e.g., glue flaps for folding carton and label printing), then it may require cutting out coater blankets. This can be done on press after the printed image is registered and in position on the sheet. However, this process can be time-consuming. One alternative is cutting the blanket off-line using a clear plastic position sheet (mylar or die vinyl). Another alternative for more precise coating requirements is to have the blanket cut by a laser die company using a CAD system laser. These alternatives can be very cost-effective compared to the time required to cut the blanket on press. For very long runs, relief plates can also be used. These plates are more expensive and should be handled with care when placed in storage.

The First Pull

The goal of any job is for the first sheet to be a good, saleable sheet with regard to register and color. If the first sheet is not in register, it must be determined why it is not.

If the following conditions are met, the goal of a good first sheet is not unreasonable:
- Plates made and mounted correctly
- Cylinders set in zero positions
- Ink fountain keys set and rollers inked up properly
- Sheet position set properly

Matching Specifications for Color and Register

A press that is well maintained, set to the manufacturer's and supplier's specifications, and using compatible materials should be able to match a correctly made off-press proof. The proof must be a final proof, approved by the customer. The proof must be clean and absent of any notes or changes scribbled on it. Matching job specifications should be straightforward and take little time.

The press operator and supervisor, CSR, salesperson, and/or the customer should make sure the images and color breaks are correct. The color match must be checked against the proof and/or ink match prints. The print quality of the sheet must be examined for anything that would render the job unacceptable to the customer.

OK to Run

The press approval should be given by the pressroom supervisor. If a customer and/or sales approval is required, then the number of people at the press should be limited. It is recommended that the approval procedure should include the press operator, pressroom supervisor, and the customer or salesperson at the press. These people should sign off on a minimum of three sheets. If press approvals are conducted in a customer viewing room, the pressroom supervisor and the press operator should be included with the customer and company managers to act as technical advisors to help expedite the approval process.

Measuring and Tracking Improvement

Makeready time is part of the cost of every job, and the cost is passed along to the customer. Keeping track of changes in makeready times is essential for helping managers to price jobs accurately. Proof of savings in a makeready time can also sometimes be used to justify purchases of new, time-saving tools or equipment.

Makeready time can be measured and graphed in various ways. One of the most obvious is average makeready time (for comparable jobs). Average makeready time can be calculated by adding up all the makeready times for a particular press over the course of, say, a week. (It is a good idea to keep separate data for each press.) Then divide by the total number of makereadies to find the average makeready time. Chart these weekly averages to determine whether the improvements are having an effect.

Makeready can also be measured as a percentage of run time. For example, if a makeready takes two hours and run time is five hours, then makeready is 40% of run time. This data should also be plotted over time on a graph.

Another way to measure makeready time is in minutes per increment. An increment can be defined as one plate or one ink change during the makeready process. For example, changing five plates and five colors of ink equals ten increments. If total makeready time for this press is 175 minutes, then each increment takes an average of 17.5 minutes to complete.

No matter what the team members select as the unit of measure, it must be charted over time, whether it be days, weeks, or months. The unit of time will probably depend on the number of jobs being printed and whether there are several shifts. It is important to remember that there is no "correct" unit of time.

As the press crew continually records makeready times on the progress chart, the plot should begin to show a downward trend.

Summary

Press makeready is a critical factor when printing a job. If the process starts on the wrong foot, it may lead to a great deal of time and cost to make it right.

Managers have considerably more control, influence, and responsibility than the press operators do for making sure that the pre-makeready concerns are handled correctly and efficiently. Everyone in the company working together as a team will help make this valuable time productive.

Video analysis is a crucial part of reducing makeready time using the SMED method. In keeping with the philosophy of continuous quality improvement, the newly improved makeready procedures should also be videotaped and analyzed to determine whether additional improvements are possible.

Another essential part of a successful attack on makeready time is to implement improvements immediately (*the three i's*). Doing so helps the makeready reduction effort to gather momentum. It also builds employee enthusiasm, morale, and acceptance of new ideas by showing employees that their suggestions are taken seriously.

Under no circumstances should anyone risk their safety (or that of other employees) by bypassing safety measures for the sake of reducing makeready time.

The makeready process outlined in this chapter is based on sheetfed press operations. However, most of the points covered will apply to web press operations.

10 Equipment Maintenance

Traditionally, the printer's equipment maintenance department and staff were considered by many as a "fix it when it breaks" operation only. Like department shift supervisors, maintenance staff are the first ones management looks for, but the last ones they want to see. Do printers show off their maintenance department when they are showing off their new, high-tech press to customers? The answer is frequently "no." The printer must now think of the maintenance department as an integral part of plant production.

The printer is faced with customer demands for much shorter production lead times, consistent high quality, and lower costs; at the same time, the printer must deal with greatly increased world-wide competition. If a printer could save half an hour on each shift of a press running three shifts, five days a week, the printer would gain almost fifty shifts on that press at the end of the year.

Equipment does not break down when it is just sitting there with the power off. Equipment failure usually seems to occur just when a hot job must run or when the customer is present for the approval. When equipment breaks down it becomes a production workflow bottleneck. When a bottleneck occurs in the production process, not only has workflow been interrupted, but so has the cashflow. If the printer's competition has less equipment failures over a period of time, the printer will probably not survive.

For the progressive printer, the maintenance team operations have become an important value-added part of the production operations. Progressive printers rely on the maintenance operations not only to fix broken equipment, but to assist in improving quality and reducing costs of wasted equipment time and materials. Equipment maintenance is starting to become a science to the progressive printer.

Unless the science of equipment maintenance is proactively established by printers, elimination of the six big losses will not be achieved.

An effective equipment maintenance program is made up of four elements: (1) restoration maintenance; (2) preventive maintenance; (3) prediction maintenance; and (4) safety. The degree of how much each maintenance element is applied may vary, depending on the nature and technology of the printer's equipment.

Restoration Maintenance

Restoration maintenance, also called repair and corrective maintenance, is the most common maintenance performed. Restoration maintenance consists of repairing a broken or damaged piece of equipment to restore necessary operating conditions. Another part of restoration maintenance includes replacing abnormal or worn parts that are causing films, plates, or production sheets to be out of specification. Restoration maintenance is basically a fix-it-when-it-breaks function, addressing sporadic and sudden equipment losses which are actually unscheduled equipment downtime. The maintenance operators become fire fighters, fixing equipment and waiting until something else breaks. There are many reasons that printing equipment breaks down:

Conventional prepress
- Vacuum-frame glass is cracked
- Plate-punch dies are bent
- Step-and-repeat machine's vacuum hose leaks
- Vacuum-frame pump fails
- Plate processor jams up
- Department power outage

Sheetfed offset presses
- Gripper-bar cam followers are worn out
- Vacuum pump is seized up
- Press sheets jam up

- Ink-roller journals are worn out
- Side guide is out of time or worn out
- Delivery gripper chains are sloppy
- Gripper bar is seized up
- Impression-cylinder drive gear is eccentric
- Double-sheet detector is not working
- Fountain solution tank's refrigeration stopped working
- IR panel is not working in auto mode
- Spraypowder application system stopped
- Electrical control system causes inconsistent stoppages of different press components
- Safety stops are not working correctly

Web offset presses
- Auto register system stopped working
- Problems with the roll splicer
- Oven temperatures are uncontrollable
- Plate-cylinder cocking mechanism is not working
- Web-break detector's sensors are not working
- Safety stops are not working correctly
- Tucker blade arm pins are broken
- Stacker/jogger is out of time
- Folder turn-bar bearings are seized up

Guillotine cutter
- Main motor drive belts are slipping
- Failure in the electrical relay
- Sheet-size sensor is burned out
- Safety system is not working
- Back jog plate does not return to starting position

Sheetfed diecutter
- Gripper bar pin is broken
- Side guide is out of time
- Feeder mechanism is worn out
- Cutter platen is not leveled
- Waste extraction is jammed

There are two types of restoration maintenance activities, emergency restoration and scheduled restoration. Emergency restoration refers to addressing a sudden and total shutdown of the equipment. The equipment cannot produce anything unless the failed component is fixed or replaced. The other, and far more important—emergency restoration maintenance—is the failure of any of the equipment's safety elements, such as safe-stop buttons and interlock guards. Immediate action must be taken by a maintenance technician or operator before the equipment can return to proper operation.

If the equipment's component failure does not shut down the machine or compromise safety or quality, scheduled restoration or maintenance is performed at a designated time, usually on weekends when the equipment is not scheduled for production.

Scheduled restoration maintenance frequently falls under two periodic activities, weekly/monthly restoration, and annual restoration. The scheduled weekly/monthly activities are usually conducted on weekends, or during periods that the equipment is usually shut down. The maintenance operations conduct repairs to equipment components that are worn out or broken. Even if the equipment is still able to operate, longer makereadies and reduced speeds are usually the result.

The key issues in weekly/monthly restoration maintenance are effectively documenting, communicating, prioritizing, and scheduling what activities must be completed each time. Effective documenting and communication for repair maintenance work to be done must include documented request forms and proper routing and review procedures. Computerized documented maintenance forms can make documenting information easier, and can make tracking equipment maintenance history very effective. The documented maintenance request or work order information should include:

Information filled in by production staff
- The name of the department and person requesting the maintenance repair
- The date of the repair request
- The requested completion date of repair
- The equipment and component failure
- Component failure: operational or safety
- Maintenance priority: safety repair, emergency repair, or scheduled repair
- If the repair work is satisfactory after completion

Information filled in by maintenance staff
- Projected completion date of repair
- The type and number of parts required and ordered
- The name of person who performed the repair

Sample of a mainte-
nance request form.
(See also Appendix.)

Maintenance Request		Priority & Date	Safety	Immediate	Scheduled	Request Number	0278-97
Department:		Equipment #				Date	
Work needed:						Date needed	
						Requested	
						Approved	
						Work	Yes
						Satisfaction	No
Maintenance Department Use Only							
Maint. Plan:						Maint. Sup.	
						Maint. Tech 1	
						Maint. Tech 2	
						Maint. Tech 3	
Type of Work	Order/Schedule	Hour Costs		TOTAL COSTS	Hours Est.		
Safety	Materials	Mat. Costs			Hours/Tech		
Immediate	Parts	Parts Costs			Hours Total		
Scheduled	Construction	Con. Costs			Completion		

- The time required to complete repair
- The costs of parts and labor
- Possible causes of component failure

Documented repair request forms can be an effective tool in managing the typical backlog of repair requests experienced by most printers. The main benefit of documented repair request forms is that they allow easier review for prioritizing and scheduling activities. The forms also enable the internal-management review process to determine true effectiveness of the maintenance program.

The tools used to perform the corrective actions are another key element in quick restoration maintenance. As basic as it may sound, tools purchased based only on price could be inadequate to effectively complete maintenance tasks. Poor-quality tools will wear out quickly and could cause damage to the equipment components that are being repaired. Accessibility of the tools is another issue when quick restoration maintenance is needed. Placing extra common tools at different equipment locations is one way to expedite emergency restorations. Mobile tool carts and portable trays will enable quick transport of all the tools required to perform emergency corrective repairs when equipment is down. Purchasing enough quality tools will greatly improve restoration maintenance efficiency.

Annual restoration maintenance is normally done once a year. The activities usually encompass restoring and changing the main components of the heavy equipment's central lubrication system, auditing a press's transfer register system, and replacing components on a predictive schedule.

These are the keys to effective restoration or repair maintenance:

- Quick repair to bring failed equipment back on line as quickly as possible
- Effective communications to properly prioritize repairs and resource utilization
- Quality repair to keep the failed equipment on line
- Quality tools procurement to help guarantee quality repairs
- Establishment and adherence to a realistic preventive maintenance program to minimize the need for restoration maintenance

Preventive Maintenance

Preventive maintenance is a program approach to preventing the sporadic and sudden failures that will totally shut down equipment. The main things that are required for a quality preventive maintenance program include knowledge of the equipment's operating components, structured scheduling, and the discipline to adhere to standards and procedures.

A key issue to optimizing a preventive maintenance program is to focus on autonomous maintenance, as discussed in chapter one. In autonomous maintenance the equipment operator's duties expand partially into the maintenance department's domain. The equipment operator's main tasks include prevention of equipment deterioration, monitoring and measuring deterioration, and assisting in restoration of equipment.

Typical activities of a preventive maintenance program will include periodic cleaning, inspection, lubrication, adjustment, and replacement of equipment components that are

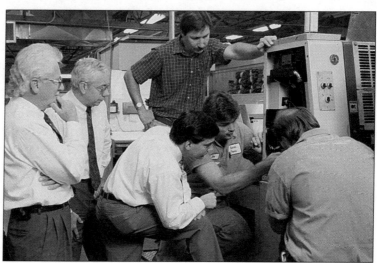

subject to high degrees of friction and repetitive operations. Common checkpoints of preventive maintenance include:

Prepress:
- All equipment safety switches
- Step-and-repeat machine pins
- Step-and-repeat machine chases, seals, and pins
- Processor's roller/brush settings and conditions
- Safe-lighting integrity
- Plate and film exposure unit lighting bulbs
- Plate and film exposure unit pump filters
- Filters in environmental system

Press:
- Safety buttons and interlock guards
- Ink and dampening rollers
- Ink and dampening roller bearings
- Transfer gripper bar cam followers
- Fountain solution recirculation tank pumps
- Double-sheet detectors
- Web turn-bar bearings
- Blanket tension bars
- Sheetfed delivery chains and gripper bars
- Press ink fountain, cleaning and maintenance
- Spray powder unit, cleaning and maintenance
- Plate/blanket cylinder bearer ring contacts and condition
- Oil in gear compartments
- Fountain solution recirculation tanks and hoses

Finishing and bindery:
- Operation light sensors
- Safety stops and switches
- Guillotine cutter drive belt, cleaning and maintenance
- Guillotine blade condition and level
- Main drive motor

Preventive maintenance activities are assigned on a periodic schedule. Normally these maintenance activities will be segregated into daily, weekly, monthly, quarterly, semiannual, and annual schedules. One of the key issues to effective preventive maintenance is adherence to the prescribed schedules of all the required activities. Typically, production management finds it too easy to postpone scheduled maintenance activities, because in their opinion, customer demands are too pressing.

That is why discipline is required by the printer to implement an effective preventive maintenance program, in turn minimizing unscheduled downtime.

Another element is tracking the failure rate and wear of parts as well as the preventive maintenance program's effectiveness, for review on a data base. Personal computers are fast becoming the most popular tool for use in effective maintenance tracking. If the printer has employees on staff who are computer savvy, the printer's organization can develop its own program from one of the popular spreadsheet programs on the market. Another option is to purchase a software program specifically developed for tracking maintenance activities. There are a few software companies that market both generic software for maintenance programs and customized maintenance programs for specific companies. The generic programs normally allow easier tracking of the printer's equipment, from copy machines to materials handling, prepress, press, and finishing/bindery equipment. A good system should be able to manage restoration, preventive, and corrective maintenance activities. The computerized maintenance program should have features that will allow the following:

- Program and produce preventive maintenance schedules and activities
- Produce and document all restoration maintenance work orders
- Track all equipment repairs made
- Produce maintenance procedures and work instructions
- Maintain a spare-parts inventory
- Issue purchase requisitions for spare parts, tools, and equipment components
- Keep a repair history database

Preventive maintenance programs are designed to minimize unscheduled downtime and equipment breakdowns. Over time the preventive maintenance program forms an equipment-repair-history database. The repair database will become one of the printer's most effective tools in eliminating both mechanical and electrical failures. If press transfer gripper cam followers experience failure every six to nine months, a repair-history database would reveal the failures and enable scheduling for part replacement every five months, eliminating unscheduled downtime of that component. If a six-color press is scheduled for twelve hours of maintenance twice a month, all the designated maintenance activities must be completed religiously. The printers who have acquired the discipline to establish and properly manage an effective preventive maintenance program have found that the increased maintenance hours have resulted in overall less downtime hours.

Prediction Maintenance

Prediction maintenance takes preventive maintenance to a higher level. Prediction maintenance utilizes more state-of-the-art technology to predict when equipment components will need maintenance before they fail. Major maintenance and overhaul intervals are now being determined by scientific methods and accurate data analysis. Prediction maintenance requires monitoring specific elements that could cause catastrophic failure of the equipment. Prediction maintenance monitoring activities include:

Lubricant analysis. This analysis includes periodic evaluation of oil lubricant reservoirs, gear boxes, and side frame housings. The analysis examines changes in color from excess heat or dirt, presence of water droplets, and presence of metal particles. The analysis includes visual and microscopic inspection every month of oil samples taken from the equipment. Oil samples can be sent to outside testing labs every three months.

Vibration analysis. Vibration analysis is used to predict failure of equipment's main bearings, and monitor the integrity of heavy equipment's setting with the floor foundation. A sensitive probe touches the locations of possible vibration. The probe can reveal the amount of vibration, compared to a normal baseline vibration, to determine if abnormalities exist. The equipment can be expensive, and the analysis can normally be conducted by outside sources for quick and accurate data.

Thermal analysis. Thermal analysis attempts to reveal any abnormal increases in temperature of various equipment components. An infrared pyrometer is a rather simple tool that will quickly measure the temperature from a remote distance. Infrared pyrometers can measure temperatures of press fountain solution in the pans, drive motors, equipment side frames, gear boxes, bearing housings, and electrical drives. The measured temperatures are compared to baseline temperatures. Increased temperatures will usually mean eminent failure of the component measured.

Crack detection. Check for cracks in side frames, shafts, pump impellers, and journals using a technique known as Magni-flux. Early discovery of metal fatigue cracks can result in more simple repairs

Noise monitoring. Periodic measurement of the noise levels of equipment components is another way to determine the existence of abnormalities. Simple listening can be part of the monitoring; however, if the area being monitored normally requires hearing protection, this way should be avoided. There are devices that will measure noise levels in decibels, so a baseline number can be established for comparison and determining if component abnormalities are occurring.

Using all of the prediction maintenance methods described above will bring the preventive maintenance program to the next level of effectiveness.

Maintenance Safety

Safety is the most important issue for consideration when addressing printing-plant maintenance effectiveness. Establishing the proper safety elements of a maintenance program should include education in safety and required protective gear, training in the use of MSDS sheets, lockout/tagout procedures, developing the proper lifting techniques, and establishing documented checklists. The safety checklists should cover extension cords, appliances, drive motors, electrical control panels, maintenance shop power tools and equipment, portable electrical tools, hand tools, ladders, compressors, compressed gases, and solvent safety containers. Poor safety practices can cause serious, possibly fatal, personal injuries, and damage to equipment.

Equipment Components Maintenance

Lubrication and Cleaning

When printers are planning maintenance, the first thing they think about is lubrication. Lubrication of moving equipment components and parts helps to prevent friction, and in turn to prevent the parts' deterioration. The primary functions of lubricants are to minimize friction, control wear, maintain normal temperatures, dampen shock, and remove contaminants. Lubricants minimize friction with higher-viscosity film on the moving parts, preventing energy loss from friction. The right lubricant will cause the equipment to run smoother and faster, and will prevent component wear and deterioration.

Accelerated component wear has three causes: abrasion, corrosion, and metal contact with metal. Poor lubrication can cause dirt and spray powder contamination, or particles breaking off of moving parts. Good lubricants helps prevent corrosion by applying and maintaining a protective film, preventing oxidation of metal parts. If metal-to-metal contact occurs due to poor lubrication, the result will be greatly increased friction and heat. Severe damage to the metal parts will occur quickly once the metal-to-metal contact is made.

Controlling component temperature is another important task of lubricants. Lubricants will act as a conductor for the heat caused by the moving parts. Special lubricants are needed to resist deterioration in high-temperature locations. High-heat drying applications such as infrared, ultraviolet, and high-velocity hot-air drying units can generate temperatures that would cause regular lubricants to break down and lose their protective abilities.

Contamination removal is another task that lubricants must do. Fresh grease will force old grease out of a component, carrying dirt and contaminants with it. Grease can also act as a seal, helping to prevent dirt from entering bearings and causing corrosion and damage. Oil will carry dirt and contaminants away from the moving parts, to be removed by filters. Periodic over-greasing and frequently changing oil filters will purge the system of contaminant, helping to increase equipment life.

Dampening shock is another function that lubricants take on. Impression-cylinder shock absorbers on high-speed sheetfed presses can build up heat with frequent feeder trip-offs at high speeds. The lubricants help to cushion the blow that the shock absorber experiences.

Cleaning is the systematic removal of dirt, contaminants, and rust from equipment components. The importance of proper cleaning cannot be overemphasized. During the greasing phase of equipment lubrication, the old and excess grease must be completely removed. This will remove from the press any dirt in the grease, and will eliminate the problem of grease contaminating the printed sheets when transported through the press. Thoroughly cleaning the side frames of equipment during maintenance activities helps eliminate hard dirt, ink, and chemical particles that contaminate other parts of the equipment.

Proper lubrication is an extremely important element of effective equipment maintenance. Consulting with the equipment's manufacturer to identify the correct lubricants and proper procedures to use is imperative to maximizing equipment effectiveness. Realistic procurement of lubricants, their storage, handling, and application will mainly determine if the printer's equipment will have a long and productive life cycle.

Electrical Components Maintenance

The majority of printer's equipment is electrically powered. Efficient electrical operation depends initially on two issues: qualified electrical technicians and correct installation of the electrical components. Qualified electrical technicians must be properly educated, have experience, and maintain on-going training. Certification from an accredited organization verifies that the electrical technician satisfied all education and training requirements. The correct installation of electrical components requires timely consultation with the equipment's manufacturer prior to and during the installation

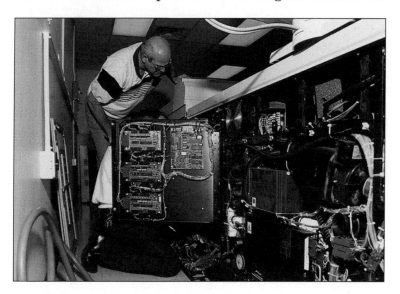

process. Bad installation of an electrical component could result in safety hazards, the component's failure, and negate any manufacturer guarantees and warranties.

Electrical problems in the printing industry are a frequent occurrence. Overcoming frequent electrical problems can come down to properly addressing a few items. In printing plants, one frequently overlooked issue is voltage fluctuation. Voltage fluctuation can cause many problems from prepress through to press. Some examples include problems with digital settings on remote-control consoles and dampener systems. In prepress areas, voltage fluctuation can cause inconsistent exposure on plate vacuum frames and step-and-repeat equipment. Voltage fluctuation has caused computer hard drives to crash on step-and-repeat equipment, resulting in total loss of operation memory. Overcoming voltage fluctuation can be accomplished by installing a voltage meter with an audio-transformer. However, many of the new exposure units today have built-in light integrators.

Another issue that the printing industry must continue to address is correct electrical wiring and grounding. The most important reason for proper grounding and correct wiring is personnel safety. Other important reasons for proper grounding and correct wiring are to maintain efficient equipment operation and to minimize equipment failure and damage. Adhering to current local building and wiring codes should be sufficient to prevent safety hazards and eliminate electrical failures. Correct wiring planning is very important to the printer's future growth and equipment expansion. When installed or expanded, the wiring system capabilities should be able to exceed any future electrical equipment requirements.

When purchasing a new building or installing new equipment, the most important things to consider are power requirements and power availability. Never assume that the required power in a new facility is available from the power company. Consult with the power company representatives prior to purchase or leasing. If the power must be upgraded to meet the printer's needs, a firm scheduled date must be established in writing. New equipment purchases must include consulting with the manufacturer on the equipment's power requirements. At times, printers purchasing new equipment have found that voltage requirements in the building don't match that of the equipment. Even if the power company can bring new power lines to the facility, the cost can be extremely high. Analyzing the power requirements for new facilities and equipment must be the number one item for consideration between the printer and the manufacturer.

Installation of new equipment must include environmental considerations and their controls. Excessively high temperatures and dirt and dust contamination can cause problems for electrical components. The majority of electrical components on newer equipment should be kept at temperatures not exceeding the manufacturer's recommendations. Dirt and dust will act as insulators, causing abnormally high temperatures to occur in electrical components. Considerations must be made to control the proper temperatures and minimize dirt contamination.

Electrical drive and pump motors are the main electrical components that the printer must properly maintain in order to maximize equipment effectiveness. Drive motors and their components take in electrical energy and convert it to mechanical drive energy. The drive and pump motors must not only be maintained, but they must be protected as well, which includes some type of enclosure. There are various types of protective enclosure systems:

Fire- and explosion-proof systems. Explosion-proof motors are designed to prevent ignition of pressroom solvent vapors and gases. The enclosure is also designed to withstand and protect from internal explosions.

Fan-cooled systems. Totally enclosed fan-cooled motors are becoming very popular. The motor is completely enclosed with either a fan attached to the motor's shaft or a separately powered fan, circulating air over the motor. These type of motors are very reliable and do not require much maintenance.

Climate-protection systems. The motor is almost completely enclosed, with air drawn up from the lower side into an area where the motor can draw the air it requires to operate.

Spill and splash-guard systems. Splash-proof enclosures prevent solid particles and liquid drops from being deflected into the motor.

Drip-on protection systems. These are normally used in relatively clean locations such as prepress areas where no water or chemicals are being used. This system prevents dirt and liquid from dropping vertically and contaminating the motor. The drip-on protection system is the lowest-cost protection system.

Although printers seem to experience many problems with electrical drive motors, in actuality only a few problems really occur. The problems that seem to occur with the highest frequency are sparking brushes, excessive noise levels, and hot bearings.

Only qualified electrical technicians should perform any maintenance work on electrical drive motors. An inexperienced press operator or an unqualified technician improperly dismantling a motor could cause major damage to the internal components. Because many drive and pump motors require three-phase electrical power wiring, a wrong connection could cause serious personnel injuries and seriously damage the motor.

Sparking brushes could be caused by a number of items, which should be checked. High degrees of sparking will probably require coil replacement.

- Check that the loads from the generator don't exceed the output rating
- Check that the drive belts aren't too tight
- Check for electrical overload with an ammeter
- Check all connections
- Check that the brushes contact the commutator evenly
- Check the condition of the brushes

- Check the commutator's condition; it should be smooth, clean, and close to the original color (darker brown)
- Check and adjust brush springs
- Check the circuit coil temperature after a short time of operation

The noise that a motor makes is normal; however, excessive and abnormal noises usually indicate probable problems to the motor's components. Measuring the decibels and knowing the noise levels of new motors and pumps would be the best ways of establishing benchmarks. Rattling sounds will usually indicate loose internal motor parts; a loose bolt may be the only problem. Squeaking noises can be caused by poor brush contact and loose spring tension. New brushes can also cause squeaking noises; they may need a little time to form better to the armature. Bumping noises are normally caused by the armature meeting resistance as it turns within the bearing. If resetting the collar to give more end play doesn't eliminate the noise, the worn bearing will require replacement. If rubbing noises occur, it is usually the result of the armature rubbing on the pole's faces; loose armature windings or worn bearings are likely causes. Monitoring the drive motor or pump sounds and noise levels can determine when maintenance will be required, before it completely fails and shuts down.

Checking the temperature of the motor or pump housing and bearings is another way to determine if repair maintenance is required. Unusually high bearing temperatures indicate abnormal bearing conditions. If the bearing temperature is too hot to touch, or if a burnt lubricant smell exists, there is a major problem with the main motor bearings. The numerous causes of this problem include:
- Dirt contamination in the bearings
- Motor and bearings are poorly lubricated
- Bearing is too tight
- Motor is out of alignment with equipment's drive section
- Shaft is crooked or bent
- No end play in motor's shaft
- Commutator temperature is too high, heat is transferring through shaft and bearing

The key to extending drive motor or pump life is to perform preventive maintenance religiously, especially cleaning and lubrication, and daily monitor the motor and pump for unusual sounds and noises.

Electrical systems must be properly maintained. Qualified electrical technicians must be the only people to work on the electrical systems, in order to ensure safety and help eliminate electrical component failure.

Mechanical Systems Maintenance

Mechanical drive systems are mechanisms that convert and transmit physical power to printing equipment. The most common mechanical drive mechanisms include chains, sprockets, gears, cams, pulleys, and belts. These mechanical drive mechanisms transmit drive power, convert and reduce speed, and physically convey and elevate materials. Virtually all printing equipment uses mechanical drive mechanisms. Proper maintenance of these mechanisms will make the difference between maximizing equipment effectiveness or mechanical failure of the equipment.

Chains. Chain drives are literally a non-stop series of chain links, the openings of which mesh with the teeth of wheel sprockets. The wheel sprockets are keyed to shafts that are connected to the drive mechanism; this allows the sprocket to both drive the chains and navigate around different spots in the equipment. There are various designs of chain drives with different purposes and strengths. Chain drives are found on:
- Feeder and delivery platforms on sheetfed presses, bindery folders, diecutting presses, and paper roller sheeters
- Delivery gripper bar systems on sheetfed presses and diecutters
- Sheet transfer carriage bar systems on certain sheetfed presses
- Web-press hoisting mechanisms for roll installation and printing-unit cassette changes
- Continuous-flow press dampening systems

Chain drive/sprocket systems have built-in advantages and disadvantages over other mechanical drive systems. The disadvantages of chain drive/sprocket mechanisms include:
- Can be noisy, and are more cumbersome than cams and gears
- Run at lower speeds than cams and gears
- Require frequent lubrication and cleaning
- More subject to failure, due to the many working parts

The advantages of chain drive/sprocket mechanisms, however, include:

- Very high drive efficiency
- Lower loads on bearings
- Drive speed remains uniform, even when there is some loss of chain tension
- Drive chain and sprocket types are standardized throughout industry, so replacements are rather easy to find
- The entire chain does not require replacement; individual links can be quickly replaced on the equipment
- Chains have a long life due to low wear and very little deterioration

Lubrication of chain drives is extremely important for long life. Application of oil is determined by the drive shaft location and the chain's speed.

- Power lubrication from oil pumps is used for high-horsepower drives
- Oil bath or splash lubrication systems require sealed enclosed housings. For moderate-speed motors, oil must be changed periodically
- Drip lubrication is a semiautomatic system that applies a small amount of oil over a predetermined time span
- Manual lubrication (oil-can application) is usually performed daily

Preventive maintenance of chain drives and sprockets is not difficult. However, realistic maintenance must be performed consistently.

- Chains and sprockets must be properly aligned
- Avoid using small sprockets, their higher speeds can wear out chains faster
- Periodically inspect sprockets for proper attachment to the shafts
- Maintain proper chain tension
- Consistently perform required lubrication of chains
- Regularly clean dirt and powder from chains

Gears. Gears have teeth formed or machined into them for tight-fitting with other gears. Gears attached to drive shafts transmit mechanical rotation power from one gear to another gear, driving each shaft. Gears have various functions, including decreasing speeds, improving drive power, changing torque, and enabling one electrical power drive motor to power many smaller equipment components. Gear types include spur gears, helical gears, spiral gears, and worm gears.

Spur gear's teeth are cut parallel to the axis of the shaft it is attached to. Spur gears can be found on lithographic dampening- and inking-system drives and bindery folders, to name a few. Spur gears are made from a variety of materials, including cast steel and iron, carbon steel, plastic, nylon, and Teflon. The plastic synthetic materials are quieter and generate less heat, but they are not as strong as metal gears. Spur gear's characteristics include:

- Most common gear used
- Low-cost and easily purchased
- Transmit power between parallel shafts
- Used on moderate-speed drives
- Have no end thrust
- Easy to maintain

Helical gears are about the most accurate, strongest, and quietest gear used in printing equipment today. The teeth are cut at an angle which causes only part of the teeth to mesh or make contact at any one time. This design allows constant drive power and eliminates the typical shock experienced from spur type gears. Helical gears are normally used in press printing and sheet transfer cylinder drives. Helical gear characteristics include:

- Teeth cut at an angle across the face of the gear
- Several gear teeth mesh simultaneously at various stages giving consistent contact
- Less vibration, less noise, and increased drive capacity
- Using bearings accommodates end thrust
- Capable of higher operating speeds

Bevel gears have teeth cut to accommodate power transmission to shafts that are running at different angles. Bevel gears are a spur gear cut at an angle. The bevel gear is about the oldest type gear used for angled-shaft power transmission. Bevel gears are frequently used in gear boxes for feeder timing to press operation. Characteristics include:

- Can transmit as much power as most helical gears
- Possess less precision than good spur gears
- Gear set is made in matching pairs, due to the difficulty in tooth control and measurement

A worm gear is a shaft that has been machined into a long spiral gear. The worm gear meshes with a helical gear on a parallel shaft or plane. Worm gears can be found on the adjustments for plate, blanket, and impression-cylinder pressure controls on nonpneumatic-operated presses. Horizontal- and vertical- bed platemaking step-and-repeat machines also use worm gears for precision register movement. Worm gear characteristics include:
- Vibration free and very quiet
- High load capacity
- Large ratio reduction in a small area
- High precision
- Generates heat and reduces efficiency at higher speeds

Proper maintenance of gears depends on their use. Gears that have high levels of power load, or which run constantly, require constant lubrication and periodic inspection. Since gears are used in the drive transmission of almost every major equipment the printer uses, proper maintenance consultation with the manufacturer is imperative.

Cams. Cams are designed and shaped to allow an equipment's components to function in precise operations. Cams either rotate, move back and forth, or remain stationary to produce the designated movement by contacting a cam follower. The cam follower can be either a round turning bearing, a non-moving flat-faced surface, or a knife edged follower. It is attached to a movable equipment element. The cam precedes the cam follower during operation, allowing the component it is attached to move at a prescribed time. Cams are used to control precise sheet transfer on sheetfed presses, ink-system ductor-roller timing, and feeder system mechanisms. Cams and followers have these characteristics:
- Can be expensive
- Have limited speed capabilities
- Are precision components
- Require minimal maintenance
- Can have a long operating life span
- Cams give dependable operation
- Can have high load carrying capabilities
- Should be frequently lubricated
- Followers contaminated with dirt or water should be replaced

- Can frequently fail due to water contamination and high heat generated by infrared or ultraviolet curing systems

Cams and followers must be properly lubricated and inspected periodically to determine if replacement is necessary. One of the more common causes for sheetfed-press image doubling is gripper system cam follower failure.

Belts and pulleys. Belts and pulleys are normally the main power transmission from the electrical drive motor to large equipment. The belt is a continuous band made up of flexible material that connects the shaft of the electrical drive motor to the main drive shaft of the machine. The pulleys are wheels pinned to the shafts of the drive motor and machine drive, then turned by the belts that are fitted onto special grooves in the pulleys. The two frequently-used belts are the flat belt and the V-belt. Flat belts are connected by glue or pins to form a continuous band. Flat belts are normally found on sheetfed-press feed tables, web delivery stackers, and delivery conveyors. Flat-belt maintenance should include:
- Make up belts ahead of time for more accurate measurements and proper glue curing
- Maintain proper belt tension
- Inspect pulley operation, check alignment, periodically clean, and lubricate
- Loosen pulley tension when changing belts

The most expensive type of belts are the V-belts. V-belts are used where heavy load carrying and high reduction ratios are present. Belt maintenance for V-belts should include:
- Inspect belt tension periodically and maintain proper tension
- Inspect belt condition and look for signs of cracking
- Never inspect belts when the drive motor is running
- Do not use lubricants or belt dressing on belts, even if a colorful salesman says so
- Replace multiple-belt systems as a set, not one component (belt) at a time
- Loosen shaft position to install belts—prying belts onto pulleys will stretch or damage the belts

The main disadvantage of a belt-drive system is the problem of belt slippage. A certain amount of belt slippage will prevent damage to equipment and to the drive motor, in the event of a paper jam-up or a component seizure. However, excessive belt slippage can cause damage to the belts and power drive loss. To help minimize belt slippage problems, certain factors should be addressed:

- Maintain proper belt tension
- Frequently listen for abnormal sounds from belt and pulley locations
- Use the right type of belt for pulleys and drive power loading
- Periodically inspect and replace belts
- Maintain proper pulley alignment
- Be aware of environmental and atmospheric changes in the equipment's location

Bearings. Bearings are a machined component into which a shaft from another component fits, and either turns or slides. The bearing reduces friction through radial circular motion and/or axial-thrust motion. Radial motion reduces friction by spinning around, not allowing the shaft to make direct contact with stationary parts of the equipment. Thrust motion allows the shaft to slide up/down or forward/backward.

Journal bearings are about the most common sliding thrust bearing. Journals, which can be made of heat-resistant metal alloys, brass, or plastic, closely fit into a side-frame housing. A thin film of lubrication helps separate the journal from the housing, reducing friction and heat. The shaft slides freely within the journal, allowing the designed oscillation.

Antifriction bearings separate the shaft from the stationary part of the equipment the same as do the journal bearings. One of the major differences of the antifriction bearings is that it is also a moving, multipart bearing. The other difference is that the bearing must both move and carry the load of the shaft. There are three types of antifriction bearings: ball bearings, roller bearings, and needle bearings. Antifriction-bearing functions can be radial or spin motion, thrust motion, or a combination of both. Antifriction bearings significantly reduce friction compared to journal bearings, due to the moving internal balls or rollers.

Ball bearings are actually journals turning on steel balls that spin within a channel or track. Ball bearings can be either a single or double row of balls. Self-aligning bearings are a double row of balls that can align when shaft deflection occurs. Roller bearings have steel rollers instead of steel balls. Roller bearings can carry more load because the rollers make more contact to the journal than balls do. Needle bearings function on thin, needle-type rollers. The needle rollers in this bearing make more contact to the journal than roller bearings, which increases the amount of friction. However, needle bearings have a smaller diameter than ball and roller bearings, so they can be used in areas with limited space. With fewer contact points, antifriction bearings overall function with lower amounts of friction than journals and bushings.

Even though antifriction bearings generate less friction than journals, they still require good lubrication. The contact points in antifriction bearings carry high degrees of load pressure. Many antifriction bearings require specific types of lubricants for long effective life. Ink-roller bearings and roller-bearing cam followers frequently fail due to poor maintenance and lubrication practices. Bearing lubrication functions in the following ways:
- Reduces friction at ball and roller contact points
- Minimizes friction heat
- Protects metal surfaces and prevents corrosion
- Helps prevent contamination from dirt and dust

Ink and dampening roller companies have found that bearing failures are a common cause of streaks and dampening control problems. The roller companies discovered that new bearings were being damaged when printers improperly installed them with metal hammers, by forcing the bearings on the shafts, by pounding the roller shafts on concrete floors, and by crookedly installing the bearings on the shafts.

Many roller companies now will install the right bearings on the roller shafts with a special press, only charging for the extra cost of the bearing. The roller companies can then truly guarantee the roller's performance.

Pneumatic Systems Maintenance

Pneumatic systems are compressed air or gases that create the power to engage an equipment's components for operation. As printing equipment technology progresses, the expansion of pneumatic systems has greatly increased. Pneumatic system compressors decrease the air volume in an enclosed area, thus increasing the air pressure. There are two types of pneumatic compressors, reciprocating and rotary.

Reciprocating compressors use a piston to compress the air within a chamber. An outlet valve releases the compressed air through a small opening at short controlled intervals. One disadvantage of reciprocating compressors is that oil and lubricants from the unit can enter the air. Larger plant compressors can create water and introduce it into the air, causing staining of paper on press and corrosion within the pneumatic components on the equipment.

Rotary compressors compress air with a rotating unit that forces air through a restricted outlet. A very common rotary compressor uses the centrifugal force of high-velocity rotating impellers forcing air out of an opening, causing temporary increased air pressure. Pneumatic systems can be found throughout a printing operation.

Pneumatic system maintenance is an essential part of maximizing equipment effectiveness. There are specific maintenance functions that should be performed periodically on compressors:

- Lockout power to compressor when performing maintenance
- Daily listen for abnormal noises
- Daily check for abnormal vibrations
- Daily check oil levels
- Daily open and drain water condensation valves
- Clean or replace air filters every week
- Check safety valves every week
- Check relief valve function every week
- Monthly inspect compressor system and hoses for leaks
- Monthly change compressors oil, inspect removed oil for contamination
- Monthly inspect main compressor system and lines for rust and corrosion
- Monthly measure and document compressor noise level decibels

Maintenance Training

The only way a maintenance program can be effective is by having people who possess the knowledge and skills in all required maintenance disciplines. Printers try to hire people who already possess maintenance knowledge and experience. One problem with hiring someone with experience is that today the number of skilled people is small. One of the usual ways printers expand maintenance staff's knowledge is on-the-job training (OJT). OJT requires someone with competent maintenance knowledge and skills to train other people.

The first step in OJT is determining if the experienced maintenance person doing the training has basic teaching skills. If basic teaching skills are lacking, there are basic "train-the-trainer" courses available, presented by local graphic arts colleges and foundations. A training curriculum should be developed, listing the basic steps that should be followed by the trainer:

1. Trainer explains the tasks being performed
2. Trainee watches and listens to the trainer
3. Trainee must explain task to trainer
4. Trainee performs task as trainer observes
5. Trainee explains steps of task performed to trainer
6. If trainee required help with performing maintenance task, go back to step number 1

The checklist should be dated and signed by both the trainer and trainee, then put in a training file. Even after the trainee has completed most of the maintenance tasks requirements, periodic reviews should be done a couple times a year. The best time to do on-the-job training is to schedule the different tasks during monthly and quarterly equipment maintenance.

Manufacturer-assisted training (MAT) is starting to become a reality in the printing industry. Manufacturers are starting to train printer's maintenance staffs in the proper preventive and repair maintenance for their equipment. Sometimes the training is conducted at the manufacturer's location; however, the training is usually done at the printer's plant. For an MAT session to be effective, there should be an agenda with a series of steps that should be followed. The first step is to consult with the manufacturer's or outside maintenance organization's servicing department. Everything the printer requires should be spelled out to the servicing people before agreeing to the servicing work.

• Communicate with the service technician before he/she arrives about the required maintenance work and the scope of the training needed
• On arrival, a formal introductory and agenda meeting should be held (the agenda should be prepared in advance)
• Everyone must understand that the primary goal of the visit is to service or repair the equipment, and the secondary goal is achieving maintenance training requirements

- A wrap-up meeting should be held at the completion of the maintenance work, to review if the work has been completed and if training requirements have been met
- Videotaping the service work is another tool for future reference. However, approval of videotaping should be secured from the manufacturer prior to the service technician's arrival.

The most comprehensive form of maintenance training is a formal apprenticeship program. A certifiable apprenticeship must include specific hours of OJT and classroom study in each maintenance knowledge requirement. The U.S Department of Labor publishes an officially recognized program. The table on the following page shows the requirements for the Machine Repair and Maintenance apprenticeship.

Maintenance Out-Sourcing

There are maintenance companies that printers can pay on a monthly retainer fee. These companies will make or acquire copies of all the engineering drawings of a printer's equipment. These companies can provide required manpower, supervision, tooling, equipment, and assistance to establish a preventive maintenance program for the printer. Maintenance companies will usually possess expensive diagnostic equipment enabling the printer to expand its program to include prediction maintenance. There are advantages as well as disadvantages to contracting maintenance activities to an outside company.

The disadvantages include:

- Reliability of a company if it is having financial problems
- Outside servicing can be more expensive each time it is performed
- There could be a long delay waiting for the service technician to arrive to repair an equipment failure
- Internal disagreements over outside people doing maintenance work

The advantages include:

- Insurance bonding by the maintenance provider gives good control of the work provided
- Expensive maintenance tools and equipment are not needed by the printer
- The printer does not have to keep updating maintenance equipment technology
- Opportunity to cancel work over poor performance is possible

U.S. Department of
Labor, Bureau of
Apprenticeship &
Training requirements

Machine Repair and Maintenance Apprenticeship

On-the-Job Training	Hours
General bench check and machine repair	750
Hand and power shop tools	400
Mechanical systems and troubleshooting	950
Pneumatic systems and troubleshooting	400
Hydraulic systems and troubleshooting	950
Electrical systems and troubleshooting	950
Material handling and storage equipment	400
Production equipment maintenance	1000
Lubrication and preventive maintenance inspection	850
Building, utilities, and HVAC	550
Shop fabrication of parts	400
Equipment installation, checkout, and safety procedures	400
Total	**8000**

Related classroom instruction   **144 hours/year**

    Basic electrical theory
    Basic electronics
    Basic hydraulics and pneumatics
    Basic blueprint reading
    Basic refrigeration, heating, and air conditioning
    Fundamentals of welding
    Machine shop fundamentals
    Industrial electricity

- Maintenance companies are more knowledgeable than the printer in setting up an effective preventive maintenance program
- Overall retainer fees can be less expensive than frequent repair work

There are certain issues that must be resolved before contracting with outside maintenance companies. Is the company financially solvent? Get references from the company's current contracts. Compare the costs of expanding the internal maintenance program versus contracting maintenance needs to an outside source.

Prepress Maintenance

Maintenance in prepress doesn't appear to be taken as seriously as in other plant production areas. Prepress maintenance should be addressed as much as other departments. Lost production due to prepress equipment failure is lost production in the entire plant production workflow.

Preventive maintenance in prepress starts with a thorough review of the manufacturer's operation manuals. Operation manuals will normally provide recommended maintenance schedules and procedures. It is also recommended that consultation with the manufacturer's technical representatives be done after reviewing the operation manuals. During the consultations questions should be asked, based on review of the operation manual. These questions should concern proper equipment operation, operating environment, training in proper preventive maintenance procedures, safety issues, and what the do's and don'ts are in regard to the manufacturer's warranties.

The best way to perform preventive maintenance consistently is to develop operational checklists. The checklists should be designed with the maintenance-activity steps in proper order for easy understanding. The checklists should be initialed and dated by the person performing the maintenance. Establishing maintenance checklists and procedures for each piece of equipment maximizes the preventive maintenance program, and helps accelerate training as well.

Proper safety procedures must be developed and included in the checklists. Understanding the appropriate MSDS sheets for the various chemicals is necessary so that the proper safety gear is available and used.

The first item to be addressed should be environment and housekeeping, in both conventional and digital prepress areas. The room temperature should be maintained at 68–75°F with 45% to 55% RH. Positive air pressure should be maintained within the department, and doors should be kept closed to help prevent dirt and dust from entering. Install and maintain proper safe lighting (based on manufacturer's recommendations), yellow for most plates and red for most film. The floors (tile, not carpeting) should be mopped daily. Minimize the number of storage boxes and containers in the department.

Press Maintenance

Presses do not breakdown when they are sitting with no ink on the rollers and the power off. Management has the responsibility both of developing preventive maintenance procedures and of overseeing their consistent implementation. The main reason that printers have difficulty consistently implementing preventive maintenance is the lack of discipline. Discipline is the adherence to maintenance scheduling and allotting the required time to complete maintenance activities.

Effective preventive maintenance programs must have various components for success.

1. All the press equipment's operation manuals must be stored in a maintenance library for bases of maintenance checklist development.
2. Preventive maintenance checklists must be developed in a professional team atmosphere. The team must consist of press management, maintenance management, press operators, and maintenance staff technicians. Everyone must understand each step in the checklist and why they must be performed.
3. Develop a clear and concise malfunction report form and routing procedures for the autonomous maintenance inspection activities carried out by the equipment operators.
4. Maintenance activity checklists must be based on different specified time requirements.

The following are examples for sheetfed-press related maintenance checklists. Many of the activities listed below are applicable to web-press operation with the necessary addition of roll stand/splicer, web guide, web break detectors, folder, stacker, and sheeter.

Daily Maintenance

- Document and date information log book
- Check oil levels in all gear encasements
- Check oil level of delivery-chain lubrication reservoir
- Check and clean cylinders bearers
- Check printing unit plate-to-blanket squeeze with packing gauge
- Check printing and coater blanket torque with torque wrench
- Check coater blanket height in relation to bearer with packing gauge
- Measure and chart fountain solution pH and conductivity in recirculation tanks
- Check that fountain solution levels are being maintained at mid-tank level points
- Check ink color tinting of fountain solution

- Check that all washup components are press-ready and at pre-makeready-2 locations
- Check that all tools are at pre-makeready-2 locations
- Have all sponges and buckets clean and press-ready
- Have all ink knives clean and at pre-makeready-2 locations
- Dispose of empty ink cans
- Wipe off side frames
- Maintain proper housekeeping of ink slab tables

Weekly Maintenance

- Drain, clean, and back-flush dampening system recirculation tanks, pans, and hoses
- Check settings and conditions of ink and dampening form rollers
- Lubricate dampening-system gear box
- Lubricate all weekly lubrication points
- Clean and check all impression and plate cylinders
- Clean, lubricate, and inspect all gripper bars/cam followers; infeed
- Clean, lubricate, and inspect all gripper bars/cam followers; transfer grippers
- Clean, lubricate, and inspect all gripper bars/cam followers and chain tension; delivery grippers and chains
- Clean, lubricate, and inspect feeder system; feeder mechanism, feed table and tapes, and side guide
- Refill delivery gripper-chain lubrication drip containers
- Clean and check spray powder unit
- Clean and check coater unit pump, hoses, and pressure settings
- Check and clean oil drip pans
- Check and clean scrap paper under press
- Check that all containers have proper hazard communication labels
- Inspect air lines, water hoses, and electrical lines
- Run roller cleaning compound into ink rollers, let stand over weekend, rinse off
- Clean all pump filters
- Sign and date weekly maintenance report

Monthly Maintenance

- Pull one set of ink rollers and inspect condition—check shore hardness with a durometer, replace abnormal rollers, clean ends of rollers going back into press
- Pull one unit of dampening rollers—inspect conditions, check shore hardness with durometer, replace abnormal rollers

- Remove and clean sides of one ink fountain blade
- Clean press side frame and ends of oscillating rollers of the roller units removed
- Clean and inspect all washup blades and trays
- Clean cylinder gaps and plate clamps
- Spray-clean and inspect all grippers and bars; infeed, transfer, delivery grippers
- Drain and replace all pump oil
- Remove spray powder, clean hoses and nozzles of spray powder unit
- Clean and check delivery fans
- Clean and inspect non-stop delivery mechanism
- Check press drive motor belt tension
- Disassemble, clean, lubricate, and inspect feeder mechanism
- Clean and inspect automatic blanket washer's pneumatic system, solvent hoses, and connections
- Clean and inspect one auto plate-mounting mechanism; lubrication, pneumatic, and drive systems
- Lubricate all monthly lubrication points
- Clean, lubricate, and inspect feeder and delivery pile, hoist chains and sprockets
- Check double-sheet and side guide sensors
- Sign and date monthly maintenance checklist

Semi-Annual Maintenance

- Drain and replace all main gear side frame oil reservoirs and clean filters
- Clean and inspect main components of press drive motors
- Clean and inspect feeder gear box and clutch system
- Run GATF Sheetfed Test Form; have the GATF Preucil lab conduct "Level 2" analysis

Annual Maintenance

- Run GATF Sheetfed Test Form; have the GATF Preucil lab conduct "Level 3" analysis
- Have a factory technician conduct inspection and maintenance of press drive and register transfer systems

Bindery and Finishing Maintenance

Bindery and finishing operations in an average printer can include such equipment as guillotine cutters, buckle folders, drills, stitchers, and collators. Folding carton operations will usually have converting equipment; platen diecutters, embossers, off-line coaters, and gluers/folders. Developing preventive maintenance programs in finishing operations is exactly the same as printing press operations.

Effective bindery and finishing preventive maintenance programs must also have various components for them to be successful:

1. All the finishing equipment's operation manuals must be stored in the finishing department manager's office as a maintenance library for developing maintenance checklists and procedures.
2. Preventive maintenance checklists must be developed in the same professional team atmosphere as printing. The team must consist of press management, maintenance management, finishing equipment operators, and maintenance staff technicians. Everyone must understand each step in the checklist and why they must be performed. Expanding operator's expertise and responsibilities into autonomous maintenance will work the best.
3. Develop a clear and concise malfunction report form and routing procedures for the autonomous maintenance inspection activities carried out by the equipment operators.
4. The maintenance activity checklists must be based on different specified time requirements.

The following examples are for developing related maintenance programs for a guillotine cutter:
• Check safety sensor operation
• Lubricate main pressure armature
• Replace and set cutting knife
• Replace cutting stick
• Clean and inspect back gauge movement mechanism
• Check back gauge square setting
• Check cutter clamp square to table

Summary

Effective equipment maintenance is one of the key issues to maximizing equipment operations and increasing a printer's income and cashflow. Equipment maintenance can actually move from a cost center to becoming an operational profit center. A true total production maintenance program must include:
• Equipment operator/maintenance staff teams working together toward elimination of the six big equipment losses
• Expanded operator knowledge, skills, and roles into daily and weekly autonomous maintenance activities
• Expanded maintenance staff knowledge and skills
• Equipment restored to optimal operating conditions

- Improved maintainability of current equipment
- Established maintenance and operation standards and adherence to them
- Improved maintenance management, communications, and scheduling

Appendix

The following checklists may be used as they are, or as guides for the printer who chooses to make custom checklists for a particular company or project.

- **Prepress Safety/Environmental Checklist**
- **Design Checklist**
- **Platemaking Checklist**
- **Plate Inspection Checklist**
- **Press Checklist**
- **Daily Pressroom Maintenance Checklist**
- **Weekly Pressroom Maintenance Checklist**
- **Monthly Pressroom Maintenance Checklist**
- **Maintenance Request Form**

Prepress Safety/Environmental Checklist

Report all injuries or hazards to your supervisor. Report any safety device that is inoperative or missing. Report any exposed wiring/exposed electrical terminals. Having a fully digital workflow, however, does not mean that you can disregard safety and environmental concerns. Following common sense and written procedures may often save injuries from occurring, perhaps even death. Take safety issues seriously. Do not smoke in production areas.

Conventional

General camera/stripping/platemaking areas

- ❒ Avoid wearing loose clothing or ties which could get caught in equipment
- ❒ Wear appropriate footwear
- ❒ Wear rubber gloves when working with chemistry; do not allow absorption into skin; gloves should be washed thoroughly with water after use
- ❒ Never operate any unfamiliar equipment
- ❒ Handle sharp instruments with care
- ❒ Do not carry knives without protective caps or covers
- ❒ Do not run with knives or scissors
- ❒ Handle and dispose of all chemicals carefully; review Material Safety Data Sheets (MSDS) and keep a notebook of them updated at all times
- ❒ Wear protective glasses
- ❒ Avoid splashing chemicals into eyes or inhaling fumes
- ❒ Know where the nearest eyewash station is and be familiar with its use
- ❒ Transport tools in a toolbox or carrying case
- ❒ Avoid looking directly into UV light sources
- ❒ Used razor blades should be tossed into a dedicated receptacle; do not throw exposed razor blades in the wastepaper basket
- ❒ Avoid looking or even glancing at camera lights; retina damage will result; wear eye protection
- ❒ Remove spills and waste as soon as possible
- ❒ Provide proper ventilation; keep air free from odor, dust, gases, fumes, humidity
- ❒ Floors should have non-slip surfaces

Light tables

- ❒ Never lean or sit on light table (glass could break; edges will go out of square)
- ❒ Avoid handling heavy objects over the surface of a light table which could cause breakage of the glass if dropped

Vacuum frames
☐   Never place 3D objects (scissors, etc.) into frame (glass will shatter when vacuum is applied)

Machine processors
☐   Require silver recovery units when using silver-based film and plates.

☐   Clean all residues from machine processors; flush out all tanks, hoses, valves; replace filters regularly

☐   Never spill deletion fluid onto skin; ignoring this leads to possible reproductive organ damage (warned by one manufacturer)

Digital

Disks
☐   Handle disks between 50°F and 125°F; never allow them to bake or freeze in a car

☐   Keep disks dry

☐   Do not touch any exposed part of disks (where data is written)

☐   Keep magnets away from disks

Computers
☐   Never drink liquids around computers; avoid moisture around computers

☐   Replace power cords when frayed

☐   Issue of electromagnetic emissions (see computer user's guide)

☐   Get up every so often and take a break to avoid muscle soreness and eye fatigue

☐   Purchase ergonomically friendly furniture

☐   Purchase ergonomically-designed mice for wrist comfort (carpal tunnel concerns)

Imagesetters and platesetters
☐   Do not remove the protective laser covers during operation or maintenance. The maximum accessible radiation level during operation and maintenance of a device is less than 0.39 microwatts.

☐   Never look directly at a laser beam

☐   Never expose other personnel to the beam

☐   Keep hands and long hair away from moving parts

☐   Never try to operate the system by defeating the interlocks

☐   If the system fails to operate, call for service

☐   Vent the heat and chemical fumes from plate-baking ovens

Signature_____ **Date**_____

Design Checklist

Please verify that all of the following information has been provided before beginning any work.

❏ Job name	❏ Number of pages (if applicable)
❏ Charge number and job number	❏ Number of colors (if applicable)
❏ Customer's name	❏ Bindery/finishing requirements
❏ Names of all interested parties	❏ Number of samples for customer
❏ All due dates entered	❏ Distribution and mailing plan
❏ Quantity to be printed	❏ Editing is complete, editing check-list is in job jacket

Please verify that you have provided all of the following information for each form before sending this job to the next stage of production.

❏ All forms identified	❏ Number of colors (and varnishes)
❏ Finished size for each piece	❏ Spot colors clearly identified
❏ Trim size for each piece (flat)	❏ Varnish style identified
❏ Stock chosen for each piece	❏ Finishing requirements
❏ Press used for each piece	❏ Should plates be saved?

Before you can consider this job complete, you must fill out the following checklist of quality control items. After you have completed this checklist, please sign and date it and place it inside the job jacket.

❏ The verification checklist was complete

❏ The job is on schedule, or I have informed my supervisor if it was not

❏ I have notified my supervisor of any low quantities of stock

❏ I have reviewed final design proofs for this job

❏ I have obtained the necessary signatures for final design proofs

❏ I have provided a folding dummy to production

❏ I have filled in the date I completed the work, and put my initials on the job jacket

❏ The job jacket has been passed to the next stage of production

Signature_____ Date_____

Platemaking Checklist

Please verify that all of the following information has been provided before beginning any work.

❐ Job name

❐ Charge number and job number

❐ Customer's name

❐ Names of all interested parties

❐ All due dates entered

❐ Editor's initials

❐ Designer's initials

❐ Preflight technician's initials

❐ Output technician's initials

❐ Quantity to be printed

❐ Number of pages (if applicable)

❐ Number of samples for customer

❐ Distribution and mailing plan

❐ All forms identified

❐ Finished size for each piece

❐ Trim size for each piece (flat)

❐ Imposition style

❐ Number-up

❐ Length of run

❐ Stock chosen for each piece

❐ Press used for each piece

❐ Number of colors (and varnishes)

❐ Spot colors clearly identified

❐ Varnish style identified

❐ Finishing requirements

❐ Should plates be saved?

Before you can consider this job complete, you must fill out the following checklist of quality control items. After you have completed this checklist, please sign and date it and place it inside the job jacket.

❐ The verification checklist was complete

❐ The job is on schedule, or I have informed my supervisor if it was not

❐ I have notified my supervisor of any low quantities of stock

❐ I have completed a plate inspection report for each form plated

❐ I have gathered the final design proofs, imposition proofs, and plate layout sheet for each form, and have them with the plates for that form

❐ I have provided a folding dummy to production

❐ I have filled in the date I completed the work, and put my initials on the job jacket

❐ The job jacket has been passed to the next stage of production

Signature_____ Date_____

Plate Inspection Checklist

☐   OK'd
   ☐ Blueline
   ☐ Double-Check
   ☐ Matchprint present (do not check plate without this)
☐   Stripping checklist is present, signed, reviewed, OK
☐   Punch is correct (no punch damage)
☐   All burns are correct
☐   No spots
☐   No holes
☐   No halations
☐   No broken type
☐   Diamonds present for web
☐   Color bars included
☐   Trim marks included
☐   Register marks included
☐   Guide marks included
☐   Color breaks are correct
☐   Colors are labeled correctly and clearly marked for press
☐   PMS areas correct
☐   Varnish areas correct
☐   No varnish on glue tabs
☐   Roll-off bars present
☐   Exposure is correct
☐   Content matches proof
☐   Color comparison for press present

☐   **OK to print**

Signature_____ Date_____

Press Checklist

Please verify that all of the following information has been provided before beginning any work.

- ❏ Job name
- ❏ Charge number and job number
- ❏ Customer's name
- ❏ Names of all interested parties
- ❏ All due dates entered
- ❏ Editor's initials
- ❏ Designer's initials
- ❏ Preflight technician's initials
- ❏ Output technician's initials
- ❏ Platemaker's initials
- ❏ Quantity to be printed
- ❏ Number of pages (if applicable)
- ❏ Number of samples for customer
- ❏ Distribution and mailing plan

- ❏ All forms identified
- ❏ Finished size for each piece
- ❏ Trim size for each piece (flat)
- ❏ Imposition style
- ❏ Number-up
- ❏ Length of run
- ❏ Stock chosen for each piece
- ❏ Press used for each piece
- ❏ Number of colors (and varnishes)
- ❏ Spot colors clearly identified
- ❏ Varnish style identified
- ❏ Finishing requirements
- ❏ Should plates be saved?

Before you can consider this job complete, you must fill out the following checklist of quality control items. After you have completed this checklist, please sign and date it and place it inside the job jacket.

- ❏ The verification checklist was complete
- ❏ The job is on schedule, or I have informed my supervisor if it was not
- ❏ The customer was notified of the approximate completion time for make-ready, in case customer wanted to be present during the pressrun
- ❏ The customer either gave press OK, or specifically declined to be present
- ❏ Run samples were pulled and stored in drawer
- ❏ Run length and paper used have been recorded on the job jacket
- ❏ I have notified my supervisor of any low quantities of stock
- ❏ I have filled in the date I completed the work, and put my initials on the job jacket
- ❏ The job jacket has been passed to the next stage of production

Signature_____ Date_____

Daily Pressroom Maintenance Checklist

❏ Document and date information log book
❏ Check oil levels in all gear encasements
❏ Check oil level of delivery-chain lubrication reservoir
❏ Check and clean cylinder bearers
❏ Check printing unit plate-to-blanket squeeze with packing gauge
❏ Check printing and coater blanket torque with torque wrench
❏ Check coater blanket height to bearer with packing gauge
❏ Measure and chart fountain solution pH and conductivity
  ❏ pH   _____
  ❏ conductivity   _____
❏ Check that fountain solution levels are maintained at mid-tank level points
❏ Check ink color tinting of fountain solution
❏ Check that all washup components are press-ready and at pre-makeready-2 locations
❏ Check that all tools are at pre-makeready-2 locations
❏ Have all sponges and buckets clean and press-ready
❏ Have all ink knives clean and at pre-makeready-2 locations
❏ Dispose of empty ink cans
❏ Wipe off side frames
❏ Maintain proper housekeeping of ink slab tables

Signature_____ Date_____

Weekly Pressroom Maintenance Checklist

❏ Drain, clean, and back-flush dampening system
    ❏ recirculation tanks
    ❏ pans
    ❏ hoses
❏ Check settings and conditions of ink and dampening form rollers
❏ Lubricate dampening-system gear box
❏ Lubricate all weekly lubrication points
❏ Clean and check all impression and plate cylinders
❏ Clean, lubricate, and inspect infeed
    ❏ gripper bars/cam followers
❏ Clean, lubricate, and inspect transfer grippers
    ❏ gripper bars/cam followers
❏ Clean, lubricate, and inspect delivery grippers, chains
    ❏ gripper bars/cam followers
    ❏ chain tension
❏ Clean, lubricate, and inspect feeder system
    ❏ feeder mechanism
    ❏ feed table and tapes
    ❏ side guide
❏ Refill delivery gripper-chain lubrication drip containers
❏ Clean and check spray powder unit
❏ Clean and check coater unit
    ❏ pump
    ❏ hoses
    ❏ pressure settings
❏ Check and clean oil drip pans
❏ Check and clean scrap paper under press
❏ Check that all containers have proper Hazardous Materials Identification System (HMIS) labels
❏ Inspect air lines, water hoses, and electrical lines
❏ Run roller cleaning compound into ink rollers, let stand over weekend, rinse off
❏ Clean all pump filters

Signature_____   Date_____

Monthly Pressroom Maintenance Checklist

❏   Pull one deck of ink rollers and inspect condition

shore hardness _____

abnormal rollers replaced?   YES   NO

clean ends of rollers going back into press?   YES   NO

❏   Pull one unit of dampening rollers

shore hardness (durometer reading) _____

abnormal rollers replaced?   YES   NO

❏   Remove and clean sides of one ink fountain blade

❏   Clean press side, frame

❏   Remove ends of oscillation rollers and clean

❏   Clean and inspect all washup blades and trays

❏   Clean cylinder gaps and plate clamps

❏   Spray clean and inspect all grippers and bars

    ❏   infeed

    ❏   transfer

    ❏   delivery

❏   Drain and replace all pump oil

❏   Clean spray powder unit

    ❏   remove spray powder

    ❏   clean hoses and nozzles

❏   Clean and check delivery fans

❏   Clean and inspect non-stop delivery mechanism

❏   Check press drive motor belt tension

❏   Disassemble, clean, lubricate, and inspect feeder mechanism

❏   Clean and inspect automatic blanket washer

    ❏   pneumatic system

    ❏   solvent hoses/connections

❏   Clean and inspect one auto plate-mounting mechanism

    ❏   lubrication

    ❏   pneumatic/drive systems

❏   Lubricate all monthly lubrication points

❏   Clean, lubricate, and inspect feeder and delivery pile, hoist chains and sprockets

❏   Check double-sheet and side guide sensors

Signature_____   Date_____

Maintenance Request	Priority & Date	Safety	Immediate	Scheduled	Request Number	0278-97
Department:	Equipment #				Date	
Work needed:					Date needed	
					Requested	
					Approved	
					Work	Yes
					Satisfaction	No
	Maintenance Department Use Only					
Maint. Plan:					Maint. Sup.	
					Maint. Tech 1	
					Maint. Tech 2	
					Maint. Tech 3	
Type of Work	Order/Schedule	Hour Costs		TOTAL COSTS	Hours Est.	
Safety	Materials	Mat. Costs			Hours/Tech	
Immediate	Parts	Parts Costs			Hours Total	
Scheduled	Construction	Con. Costs			Completion	

Bibliography

Apfelberg, Herschel, *Maintaining Printing Equipment.* GATF: 1984.

Bertolina, Ronald and Koehler, Charles, "Total Prepress Maintenance." *GATFWorld,* March/April 1997.

Blanchard, Kenneth and Johnson, Spencer, *The One Minute Manager.* Berkley Publishing: 1987.

Freidel, Don, "Using SPC Tools to Solve Printing Problems." *GATFWorld,* November/December 1994.

Geis, John, *Sheetfed Press Preventive Maintenance.* Technical Services Report #7230, GATF: 1982.

Handbook for Folding Carton Production. Boxboard Containers, Intertec Publishing Corp.: 1995.

Nakajima, Seiichi, *Introduction to TPM: Total Productive Maintenance.* Productivity Press: 1988.

Nakajima, Seiichi, *TPM Development Program.* Productivity Press: 1989.

Peach, Robert W. and Ritter, Diane S., *The Memory Jogger 9000.* GOAL/QPC: 1996.

Shingo, Shigeo, *Revolution in Manufacturing: The SMED System.* Productivity Press: 1985.

Stanton, Anthony, "Using the GATF Sheetfed Color Test Kit." *GATFWorld,* July/August 1989.

Strader, Joe, "Maintenance Training on the Job." *High Volume Printer,* June 1996.

Index

About the Author

Kenneth E. Rizzo, a senior consultant, technical/quality systems, at the Graphic Arts Technical Foundation, specializes in sheetfed press operation and troubleshooting. A highly qualified graphic arts professional with over 25 years of experience, Rizzo answers inquiries and conducts Technical Plant Assessments. During a TPA, Rizzo observes, examines, and evaluates the printer's production operation with an emphasis on troubleshooting the plant's methods, techniques, and equipment.

Rizzo's strengths are rooted in a background of print production management, pressroom personnel supervision and training, platemaking and press operation, environmental and safety standards and procedures, paper, ink and chemistry interactions, continuous improvement techniques, and implementation of the ISO 9000 standard.

As a technical trainer, he serves as the leader for several workshops. He participates in GATF's custom training programs which are conducted on-site in printing plants worldwide, and he lectures for technical and print production seminars. Additionally he has written technical articles for *GATFWorld* magazine, and his seminars and work have been featured in many trade magazines.

Rizzo developed production systems processes, including GATF's Quick Response Makeready program, designed to improve the printer's technical and quality systems based on the SMED system. Rizzo has adapted and developed the concept of total production maintenance specifically for the needs of today's printing industry. He introduced it to printers at the 1996 GATF/NAPL Sheetfed Pressroom Conference.

About GATF

The Graphic Arts Technical Foundation is a nonprofit, scientific, technical, and educational organization dedicated to the advancement of the graphic communications industries worldwide. Its mission is to serve the field as the leading resource for technical information and services through research and education.

For 73 years the Foundation has developed leading-edge technologies and practices for printing. GATF's staff of researchers, educators, and technical specialists partner with nearly 2,000 corporate members in over 65 countries to help them maintain their competitive edge by increasing productivity, print quality, process control, and environmental compliance, and by implementing new techniques and technologies. Through conferences, satellite symposia, workshops, consulting, technical support, laboratory services, and publications, GATF strives to advance a global graphic communications community.

The Foundation publishes books on nearly every aspect of the field; learning modules (step-by-step instruction booklets); audiovisuals (CD-ROMs, videocassettes, slides, and audiocassettes); and research and technology reports. It also publishes *GATFWorld,* a bimonthly magazine of technical articles, industry news, and reviews of specific products.

For more information on GATF products and services, please visit our website http://www.gatf.lm.com or write to us at 200 Deer Run Road, Sewickley, PA 15143-2328 (phone: 412/741-6860).

Other Products of Interest from GATF

To place an order, or for more information about any of the products and services mentioned in this book, please call GATF at 412/741-6860; fax at 412/741-2311; email at info@gatf.lm.com; or write to GATF, 200 Deer Run Road, Sewickley, PA 15143.

Press Testing

Order no.

Products with two order numbers are available in negative and positive form. (70XX—positive, 71XX—negative.)

Texts

Skill Standards for the Printing Industry

SecondSights (Reprints of relevant articles from *GATFWorld*)